A CRITICAL INTRODUCTION TO MATHEMATICS EDUCATION

The second edition of Mark Wolfmeyer's award-winning primer offers future and current math teachers an introduction to the connections that exist between mathematics and a critical orientation to education, one that accounts for race, social class, gender, sexuality, language diversity, and ability.

Expanded and updated from the first edition, this book demonstrates how elements of human diversity and intersectionality have real effects in the mathematics classroom, and prepares teachers with a more critical math education that increases accessibility and equity for all students. By refocusing math learning toward the goals of democracy and social and environmental crises, the book also introduces readers to broader contemporary school policy and reform debates and struggles, especially in light of Covid-19 and the ongoing struggle for racial equity.

Featuring concrete strategies and examples in both formal and informal educational settings, as well as discussion questions for teachers and students, text boxes with examples of critical education in practice, a glossary, and suggestions for further reading, Mark Wolfmeyer shows how critical mathematics education can be put into practice, relevant for undergraduate and graduate students in education, current teachers, and teacher educators.

Mark Wolfmeyer is an associate professor in the College of Education at Kutztown University of Pennsylvania.

Critical Introductions in Education Series

Series Editor: Kenneth J. Saltman

The Politics of Education: A Critical Introduction, 2e
Kenneth J. Saltman

Mathematics Education: A Critical Introduction
Mark Wolfmeyer

English Language Arts: A Critical Introduction
Julie Gorlewski

**A Critical Introduction to Mathematics Education: Human Diversity
and Equitable Instruction**
Mark Wolfmeyer

For more information about this series, please visit: https://www.routledge.com/
Critical-Introductions-in-Education/book-series/CRITEDU

A CRITICAL INTRODUCTION TO MATHEMATICS EDUCATION

Human Diversity and Equitable Instruction

Second Edition

Mark Wolfmeyer

Routledge
Taylor & Francis Group

NEW YORK AND LONDON

Designed cover image: © GettyImages /PeterPencil

Second edition published 2023
by Routledge
605 Third Avenue, New York, NY 10158

and by Routledge
4 Park Square, Milton Park, Abingdon, Oxon, OX14 4RN

Routledge is an imprint of the Taylor & Francis Group, an informa business

First edition published by Routledge 2017

ISBN: 978-1-032-34518-5 (hbk)
ISBN: 978-1-032-34507-9 (pbk)
ISBN: 978-1-003-32256-6 (ebk)

DOI: 10.4324/9781003322566

Typeset in Bembo
by SPi Technologies India Pvt Ltd (Straive)

For mathematics teachers everywhere

CONTENTS

PREFACE

This book is for mathematics teachers who want to teach for all learners in their classrooms. Many typically think that mathematics is objective and so straightforward that teaching it should be universal; what works for one will work for anyone. However, we have such stark differences in student performance in mathematics, and these patterns in the data fall along distinct lines among our various identities. For example, why does the data indicate that white students always outperform students of color in our classrooms? This simply cannot continue. Together we will realize that our teaching approaches are in part to blame. We cannot continue to teach our students with a one-size-fits-all approach. To perform better in the classroom, we'll need to understand more deeply about human diversity and how it relates to mathematics instruction.

We start in Chapter 1 by unsettling the notion that mathematics is objective and clearly defined. Philosophers, anthropologists, and historians all paint a messy picture of the mathematical world. It is far from simple and beautiful in its complexities and varieties. Mathematics has been developed by a broad range of people across human diversity and throughout history; mathematics comes in a variety of forms from mathematical practices embedded in cultural life to mathematical modeling that can solve challenging real-world problems to the pure mathematical inquiries full of deduction and reasoning. Mathematics is varied and diverse and as its teachers we reflect this diversity by widening what we offer students; we also approach mathematics instruction uniquely for each student to reflect their multiple identities.

In Chapters 2 and 3, we continue to frame a mathematics instruction across the facets of human diversity. First, we review the dominant perspective in mathematics education research today, typically referred to as the reform approach. Prioritizing student-centered instruction that emphasizes the "thinking" and "doing" of mathematics has provided students with greater access to all of mathematics. And yet, despite these efforts, Chapter 3 demonstrates how they failed to address marginalized identities in our classrooms. Opportunity gaps for high-quality instruction are readily displayed in the data. This forces mathematics teachers to challenge our instruction for better outcomes for our students of color, those from lower social classes, our girls and women, our students in the LGBTQ+ community, our students with disabilities, and our emergent bilingual students who come to us with languages additional to English.

In Chapters 4 through 9, we take each of these social identities in turn to explore their connections to the mathematics classroom. First, we engage the chapter's facet of human diversity (the chapters each take on: race, social class, gender, sexuality, ability, and language) by exploring the most advanced thinking on it that comes from social sciences disciplines like anthropology, sociology, and psychology. We next engage with the broader discussions of this facet with public education generally before moving to the specific research and practitioner knowledge about how it relates to the mathematics classroom. In each chapter, I end with discussion prompts, classroom tips for our teaching, and a glossary of terms to make the work ahead more tangible. Perhaps you are reading this book for a course or as part of a learning community of mathematics teachers. I encourage you to engage with the features at the end of each chapter with others who are reading along with you.

I offer one caution upfront: The structure of this book presents a dangerous implication that we can somehow treat facets of human diversity (such as race or gender) in isolation. This is not possible. We must approach the work ahead by emphasizing how individuals and groups of individuals always have layered identities, a concept that social scientists call intersectionality. You'll notice that each chapter has many references and connections to other identities beyond the chapter's particular focus on one facet of human diversity. Despite the realities of intersectionality for individuals and groups, it still makes sense to organize the book in the chapters as they are. This allows for the clear and focused discussions we need to have for each particular facet of human diversity. The intention, however, is not to separate them out as if they occur in any isolated way or to prioritize one facet of diversity over another.

Some of you may approach the work ahead with enthusiasm, having some experiences in diversity and an eagerness to see its applications to our work as mathematics teachers. I'm glad you are reading this book and learning more! Others may find a newness to approaching certain topics contained in this book and possibly some discomfort for certain chapters. I'm also excited that you are reading this book and have a suggestion for you: Chapters 4–9 do not need to be read in order, and it might make sense to read a chapter that focuses on a facet of human diversity with which you are more comfortable engaging as a positive starting point. Maybe this chapter contains material of personal significance to you. This will help to convince you that identity matters in our classrooms and can motivate you to move into other areas that are less familiar for you.

Your goal as a mathematics teacher is the greatest possible outcome for all your learners. The discussions contained herein will convince you that your learner's *identities matter significantly* as related to your ability to teach them mathematics. I have attempted a collection of the most advanced thinking on these topics, complemented by a set of best practices in our field, to help you in your efforts to teach your students. Mathematics teachers who do not engage these concepts are perpetuating classrooms and mathematics instruction as a privileged space only for the successes of white, upper class, heterosexual, able-bodied, English-speaking, cisgender men. I am excited that you are a mathematics teacher willing to consider these points carefully and move our field toward better outcomes for all learners!

ACKNOWLEDGMENTS

First and foremost, I thank members of my family for their support in writing this book. My partner Ellie Escher and children Beatrice and Guy WolfmeyerEscher have provided much encouragement and the space to write. Thanks to my parents Helen and Paul Wolfmeyer, in-laws Gus and Connie Escher, and siblings David Wolfmeyer, Beth Cocuzza, and Amy Escher.

Thank you to Kenneth Saltman for inviting me to submit a proposal for the series in its first addition and for continued support in writing this second edition. Thank you to Routledge Editor Heather Jarrow, Editorial Assistant Sophie Ganesh, and Simon Jacobs, formerly of Routledge, who helped with the project as well.

A number of others provided encouragement, support, and/or extensive feedback on this book. These include Brian Greer, Erika Bullock, Brian Lawler, Alexander Moore, Andrea McCloskey, John Lupinacci, Nataly Chesky, Vale Deeter, John Ward, Amber Pabon, Andrew Miness, Brenda Muzeta, Patricia Walsh Coates, George Sirrakos, Miriam Tager, and Zachary Stackhouse. Finally, thanks to anonymous reviewers who provided substantive feedback to augment the book's contents and accessibility.

1

WHAT IS MATHEMATICS?

Answers from mathematicians, historians, philosophers, and anthropologists

In this book, we will focus on teaching mathematics to all learners across the facets of human diversity. Before engaging concepts within human diversity and their relationships to the math classroom, we will start by examining the discipline of mathematics to reveal its own diverse history and current practice. Mathematics is among many disciplines that people have produced over our collective history. Like other content areas, mathematical knowledge and its subtopics originated for many different reasons, and the topics and nature of math have been debated and challenged over time. Many of us may tend to think of mathematics as a set of ideas that are concrete and well defined. For example, you may be asking, "Is it not objectively true that $1 + 1 = 2$ and there is no evidence to the contrary?" While most, if not all, people practicing mathematics will agree to this as mathematical fact, there are several additional questions we can pose about mathematics and its nature, many of which provide active areas of debate. We can ask, "What counts as mathematics?" or "Is there a universal mathematical knowledge that transcends time and cultures?" As with all chapters in this book, to answer these questions we will take a people-centered approach, looking to the study of mathematical knowledge by examining the people doing mathematics and thinkers who write about the history, philosophy, and anthropology of mathematics. In doing so, we will push ourselves to think of mathematics less as a static world of academic (mostly white, male, and Eurocentric) development and rather as a multicultural and social activity constantly evolving over time. As with all chapters in this book, I conclude with classroom tips and a summary table of important terms. Some readers may find it useful to flip to the end of the chapter and read these first before moving through the details in what follows.

DOI: 10.4324/9781003322566-1

An introduction to mathematical behaviors and pure and applied mathematics

We will start by looking at typical mathematical behaviors that come to mind as well as the work of academic mathematicians. I am starting us here because it is often the place with which many readers of this book will have familiarity. Note however that the subsequent sections on philosophy, history, and anthropologies of mathematics will give us a much more complete and broad sense of mathematical behavior than what we begin to think about here.

In describing a few mathematical behaviors, first, there is the mathematical behavior of computations with numbers. This is a common point of reference for an understanding of mathematics. Someone you know may say, "I am so bad at math!" when confronted with a task that requires multiplication, division, addition, or subtraction of two numbers. The "basic four computations" are likely the first thing that comes to mind when most people think of math.

Let's play in this conception of math a bit. See if you can add the numbers 781 and 312 without a calculator. Many of you might like to reach for a pencil and paper and set up the problem as you were taught in school. These pencil and paper methods are referred to as *standard algorithms* for mathematical computations. At some point, a person (like a teacher or an adult close to you) might have encouraged you to try to answer such problems without the use of the standard algorithm, instead prompting you to develop a reasoned computation strategy. Try the problem again. The goal is not to say the answer (1,093), but to argue how you arrived at the answer. Now try to give a mental computation for 43 times 15. Give yourself a chance to come up with some methods before reading the next paragraph.

There are many ways to compute the answer of 645, and here is one of these as an example. Break the 43 into 40 and 3. You know you need to multiply 40 times 15 and 3 times 15. The second one is easy by repeated addition (45). The first problem can be made simpler again by multiplying 4 by 15 (60) and adding an extra 0, because the problem is really 40 by 15 (600). Now you add your two parts together to get 645. You may be able to follow this short narrative, and/or Figure 1.1 may help. This method relies on the fact that we understand the

$$43 \times 15$$

$$40 + 3$$

$$40 \times 15 = 600$$
$$(4 \times 15 = 60)$$
$$3 \times 15 = 45$$
$$645$$

FIGURE 1.1 A worked-out mental math example

concept of multiplication. We can think about multiplication as repeated addition. For example, 5 times 4 is 5+5+5+5. This was an important part of our method. For one, we easily saw how 45 is the product of 3 and 15. It also allowed us to break apart the 43 into 40 and 3, and then add the results together.

I have been describing the knee-jerk response to the main question of this chapter: Mathematics means "doing these types of computations." Mathematics, however, is much more than this, and we need to look at the other behaviors that can be considered mathematical. Where to go next but with those whose professional activity centers on such behaviors: mathematicians. Many university math departments, where mathematicians often work, are split into two divisions: pure and applied. In *What is Mathematics, Really?* Ruben Hersh (1997) differentiates the two as follows: "Mathematics that stresses results above proof is sometimes called 'applied mathematics.' Mathematics that stresses proof above results is sometimes called 'pure mathematics'" (p. 6). Pure mathematicians work within abstract worlds to prove things that have little association to a particular physical (or social) situation. Applied mathematicians do the opposite: they start with these physical or social situations and adapt the work of pure mathematicians to address particulars within the real-world application. We will review these distinctions within mathematics more thoroughly and also note how they are not always so cut and dry as labels for academic mathematics.

Within each division are a host of topics. Among the topics in pure math are number theory, algebra, geometry, topology, calculus, analysis, and combinatorics. Generally speaking, applied mathematics includes any kind of mathematical knowledge that has made a connection to a real-world problem. Such endeavors have spawned particular fields of their own, such as differential equations, mathematical modeling, statistics, mathematical physics, and game theory. Thus, a variety of mathematical topics are at play among the work of mathematicians. The distinction between pure and applied mathematics proves highly relevant as we look at mathematics critically in order to conceptualize how we will teach it and for what purpose. I want to illuminate this distinction with two mathematical examples from basic and intermediate mathematics. First, returning to our computation problem, we can see how 43 times 15 is an abstract concept, and our method of computing it required a conceptual understanding that multiplication is repeated addition. It is not too difficult to imagine a context in which we would have to apply such knowledge. For example, I might want to determine how much compost I need to spread on my garden that has dimensions 43 feet by 15 feet. In a sense, multiplication is both pure and applied mathematics.

For another example, you may have learned a bit of trigonometry in your mathematical experiences. One fact in basic trigonometry is about the relationships between the side lengths of a right isosceles triangle, that is, a triangle with one 90 degree angle and two 45 degree angles. If you know one of the lengths of the "legs," the two sides that are equal, then you can approximate the longest side by multiplying the leg length by about 1.4 (the exact number is the square

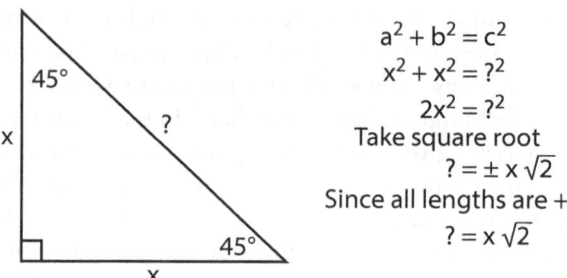

$$a^2 + b^2 = c^2$$
$$x^2 + x^2 = ?^2$$
$$2x^2 = ?^2$$
Take square root
$$? = \pm x \sqrt{2}$$
Since all lengths are +
$$? = x \sqrt{2}$$

FIGURE 1.2 An elementary example of pure mathematics

root of 2). You can prove this using the Pythagorean theorem, and this is shown in Figure 1.2 when you assign the length of the two equal sides as x. This is an elementary example of pure mathematics.

As for applied mathematics, such trigonometric relationships are readily applicable to the real world. Back to gardening, let's say you have a square garden that measures 20 feet on each side and you need to know the length from one corner to its opposite corner (the diagonal length). Using the mathematics described above, you can approximate this length as about 28 feet.

When thinking of the two broad branches of mathematics, applied mathematics may seem the less daunting of the two. The work of pure mathematics involves the invention of new material, whereas applied math takes these efforts to solve new problems. However, this perception is not at all the case, as applied mathematicians have their work cut out in dealing with the "messy" real world. To solve problems, they need to use mathematical ideas that have been created in an ideal, imaginary world. It is also true that venturing into the new frontiers of pure mathematics is highly daunting and many work tirelessly for years at this. This discussion between the two branches suggests that mathematics teaching must include both. Many attempts have been made to "make mathematics relevant" with the inclusion of applied mathematics. Some of these are more contrived (think of those textbook word problems that do not resemble real situations), and there are other examples that reflect the work of applied mathematics more accurately. As for pure mathematics, teaching mathematics in this way implies that we provide experiences where our students will come to "discover" mathematical ideas as pure mathematicians do.

For the latter, here is an example adapted from Paul Lockhart's (2009) *Mathematician's lament: How school cheats us out of our most fascinating and imaginative art form.* This helps us to explore pure mathematics a bit more with an example problem coming from the mathematical branch of number theory. The most helpful part of Lockhart's example is the fact that he encourages us to imagine any number as a pile of rocks. So, to begin: imagine the numbers 9 and 4 as two separate piles of rocks. Now consider arranging them in various ways, like a line, a circle, a square, more than one object, etc. It could be helpful for you to get out some beans or something else to use as your "rocks."

I assume you know that 9 is an odd number and 4 is an even number. Continue playing around, arranging your rocks, and this time focus on these facts about even and odd. Is there a way you can arrange the 4 to show that it's even? Maybe create a set of 6 rocks and 8 rocks as well; then you have a few even numbers to play with. Do the same for odd. Come up with as many arrangements as possible. The longer you play, the more likely you are to come up with the arrangement I hope you do. This arrangement, Figure 1.3, appears later on this page. Don't peek until you've played enough!

Any even number can be placed in two rows. In other words, we can pair up each of the rocks. If we try to place an odd number of rocks into two rows, one of the rocks is left without a pair. Fascinating!

This representation can be used to answer some interesting questions in number theory. Use these representations (numbers as rocks) to prove an answer to the following questions: What kind of a number do you get when you add two even numbers? Two odd numbers? An even and an odd number? Enjoy playing with these representations and work on a problem in pure mathematics!

There is one final note about the notions of pure and applied math. As with most dichotomies, I suggest you consider them as useful categories to further appreciate the nuances of mathematics. In doing so, however, we cannot come to understand them in any way as distinct entities. There is much of mathematics that may fall into one or the other category, and Greer (in preparation) provides a clear summary of challenges made to the notion of mathematics' "two faces," for example, in the ways that mathematical modeling as an area of academic mathematics performs a cyclical process of moving from a real-world context, next to a theoretical mathematics, back to the real-world context, and again to revising the model created in the theoretical space. As well, sometimes the notion of application can suggest that pure math always comes first and, by further logic, that pure math is somehow superior to applied math. However, Hersh (1997)

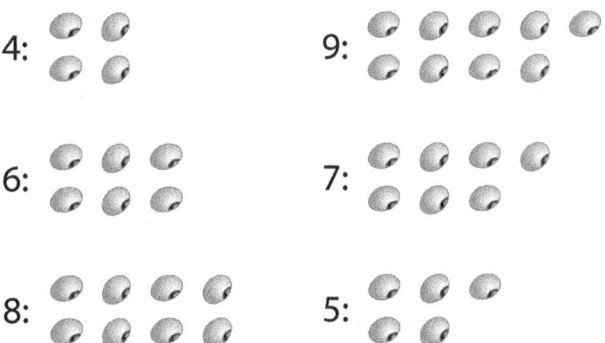

FIGURE 1.3 Beans arranged to deduce theorems from number theory

notes that pure mathematicians value applied math just as highly as pure math. Part of this is the fact that much of pure math has occurred as the result of applied math. "Not only did the same great mathematicians do both pure and applied mathematics, their pure and applied work often fertilized each other. This was explicit in Gauss and Poincaré" (p. 26).

What does the philosophy of math tell us?

Exploring these various branches of academic mathematics has begun to answer our central goal in this section. After such an introduction, it seems appropriate to next take a look at the work of philosophers of mathematics. We might suspect such work aims to answer the question in a very direct way. Instead, learning the mainstream and contrary viewpoints within the philosophy of math presents an important consideration you may not have anticipated. Namely, the following review asks us to decide whether mathematics exists outside of our having discovered it, as a set of ideals, or something that was created by people. The latter represents some of the more controversial and critical aspects to thinking about what mathematics is.

Three books prove helpful in thinking about the philosophy of mathematics: Hersh's (1997) *What is mathematics, really?* (1997), Ernest's (1991) *The philosophy of mathematics education*, and Hacking's (2014) *Why is there philosophy of mathematics at all?* They provide a significant review of the major names in philosophy of mathematics, with Hersh as a narrative style, Ernest as an in-depth and technical review of each strand of philosophies of mathematics, and Hacking with a highly detailed attention to several conventions in mathematical practice, such as his robust explication on proof in mathematics.

All give a sense of the debates in philosophies of mathematics. Hacking's title puts forth the challenge present in our chapter here: Why would we need a philosophy of mathematics if it is something pure and objective? Hacking demonstrates the distinctions within mathematical practices and its key thinkers, such as the delineation of a Cartesian versus Leibnizian style of proof, to highlight how mathematics contains nuance and varieties in practice. Similar to our discussion earlier in this chapter, Hacking reviews the myriad practices and definitions for mathematics to suggest that there really is not one clear definition, although many have attempted to do so over the years. Hersh (1997) starts with one mainstream viewpoint, "mathematics is superhuman—abstract, ideal, infallible, eternal." Several of the thinkers espousing this view are "tangled with religion and theology" (p. 92). On the other hand, those with the contrary viewpoint see mathematics as a human activity or human creation. Hersh refers to this group as the humanists, and he identifies thinkers on both sides throughout the Eurocentric history of the philosophy of mathematics. As an example of humanist mathematics philosophy and a philosopher of mathematics himself, Ernest's specific critical viewpoint is termed "social constructivism."

The mainstream view of the philosophy of mathematics is clearly exhibited by Ancient Greek philosophers, including the Pythagorean society and Plato. The Pythagorean society situated their mathematical activity within a quest for spirituality. For example:

> The Pythagorean discovery that the harmonics of music were mathematical, that harmonious tones were produced by strings whose measurements were determined by simple numerical ratios, was regarded as a religious revelation... The Pythagoreans believed that the universe in its entirety, especially the heavens, was ordered according to esoteric principles of harmony, mathematical configurations that expressed a celestial music. To understand mathematics was to have found the key to the divine creative wisdom.
>
> *(Quote attributed to Richard Tarnas in Hersh, 1997, p. 93)*

The perfect, ideal relationships witnessed in music and elsewhere indicated a harmonious truth and beauty. It was, as if to say, to lead fully spiritual lives, to become more beautiful and perfect, we as people must learn mathematics and discover such harmonies in their existence.

This is regarded as a stepping-stone toward Plato's famous notion of ideals, in which mathematics played a significant role.

> Platonism is the view that the objects of mathematics have a real, objective existence in some ideal realm. It originates with Plato, and can be discerned in the writings of the logicists Frege and Russell, and includes Cantor, Bernays (1934), Hardy (1967) and Godel (1964) among its distinguished supporters. Platonists maintain that the objects and structures of mathematics have a real existence independent of humanity, and that doing mathematics is the process of discovering their pre-existing relationships. According to Platonism mathematical knowledge consists of descriptions of these objects and the relationships and structures connecting them.
>
> *(Ernest, 1991, p. 29)*

Such Platonism is one of a few varieties in the mainstream view of the philosophy of mathematics that imagines mathematics as fixed, neutral, and value-free. Another strand of philosophy of mathematics, absolutism, includes intuitionism, formalism, and logicism. Logicism describes a standpoint in which all of mathematics can be described within logical terms and principles; formalism essentially claims mathematics to be the practice of defining mathematical truths through symbols; and intuitionism that mathematics must rely on the construction of proofs and objects (Ernest, 1991, pp. 7–12). As Ernest (1991) writes, absolutist schools of philosophy should have included accounting for the nature of mathematics, including external social and historical factors, such as the utility of mathematics, and its genesis. Because of their narrow, exclusively internal preoccupations, these schools have made no contribution to a broadly conceived account of mathematics.

(pp. 23–29)

Hersh points out that in many instances, such absolutist and Platonic frames of mind coincide with a religious or theological perspective. For example, Descartes attempts to prove that God exists because a perfect triangle exists within his mind. In this way, mathematics is seen as a set of divine ideals to be discovered by people.

On the other hand, a major branch of philosophy of mathematics is termed "fallibilism": the view that "mathematical truth is fallible and corrigible, and can never be regarded as beyond revision and correction" (Ernest, 1991, p. 18). If Platonism and absolutism rest on mathematics' attempts to discover what is indubitable, fallibilism represents a mathematical body of knowledge that we know to be true simply because we have not proven it false yet. More broadly, this description fits under what Hersh terms as a "humanist" mathematics, where mathematics is seen as the product of human interaction and development. Ultimately, any mathematical truth has been argued by people and is thus the product of such human experiences.

Among many more, two philosophers of mathematics are important here: Ludwig Wittgenstein and Imre Lakatos. Wittgenstein suggested our earlier example simply as follows: "1 plus 1 equals 2 because we have decided it so." This is an important step and a clear disagreement with the mainstream absolutist and Platonic perspectives. The sum of two 1s equals 2 not because of some ideal set of numbers existing in their perfect form and perhaps with divine intervention. Instead, over the course of human history we have decided it so. Any person is free to disagree with the equation. For example, they might say that the sum is 3. However, this person would have difficulty participating in the mainstream use of numbers that society has developed over time.

Lakatos described the process in more detail. Influenced by Karl Popper, a philosopher of science, essentially he claimed that every mathematical truth is the result of argumentation. Popper had revolutionized science by arguing that scientific theories are only guesses waiting to be disproved by experimentation. Similarly, mathematical truths are statements either proven true via an argument that is accepted by the mathematical community (a "proof") or proven false given a "counterexample" or perhaps other means. Earlier, we played with the notion of proof when attempting to explain that the sum of two even numbers is even earlier in this chapter. For the record, mathematicians would not accept our play with rocks as a formal proof, but for conversations sake this will be helpful. How about the counterclaim that two even numbers add to an odd number?

Well, that is easy to argue against with the use of a counterexample. Take a minute to give one now by playing with your rocks, beans, etc.

Proving and providing counterexamples lead to mathematical "truth." Hersh writes;

> Instead of a general system starting from first principles, Lakatos presents clashing views, arguments, and counter-arguments; instead of fossilized mathematics, mathematics grows unpredictably out of a problem and a conjecture. In the heat of debate and disagreement, a theory takes shape. Doubt gives way to certainty, then to new doubt.
>
> *(Hersh, 1997, p. 211)*

Embedded within Lakatos' assertions is the assumption that mathematics is not ideal truth, and certainly not something created by a superpower that we unearthed. This point clashes somewhat with the mainstream philosophies of math. However, I suggest it does not contradict the mathematical quest of discovering truth and beauty, such as exemplified by the Pythagoreans and their attempts to live the good life. What remains important in our objectives for studying mathematics education critically is that we examine the variety of natures of mathematics expressed by this philosophical work.

A specific option available to us is Ernest's own philosophy of mathematics, what he terms the social constructivist. Hersh positions Ernest's work within a humanist philosophy of mathematics. Ernest begins with the clear statement that mathematics is a "social construction." This concept will be utilized consistently through this book, as we look at other examples of social constructs such as race, class, and gender. For now, think about social constructs as objects that we might think of as "fixed realities" but instead have been developed over time in social settings. Social theorist Michel Foucault uses the term "regimes of truth." Do we not usually think of mathematics as fixed and objective? True and value free? And certainly the earlier philosophies of mathematics, like absolutism, reinforce this assumption. Alternatively, we can think of mathematics as something manufactured by social groups, a social construct.

To begin, Ernest orients us to a claim that mathematics is delivered via language, which on its own is a construction of the social experience. In describing the process in which mathematical knowledge comes to be, Ernest distinguishes between objective and subjective mathematical knowledge. An individual constructs subjective mathematical knowledge; objective mathematical knowledge is that which has been understood and accepted by a community of mathematical knowers. When an individual proposes a new mathematical statement, she uses language. The community of mathematical knowers uses objective mathematical knowledge to make sense of this new statement. The body of objective mathematical knowledge is the discipline of mathematics, and such a conception fits within a fallibilistic claim that everything known mathematically has simply not yet been disproven.

Ernest stops short of discussing this community of mathematical knowers. He claims that a philosophical inquiry can only go so far as to describe the generation of mathematical knowledge through this social process.

> It would be inappropriate in a philosophical account to specify any social groups or social dynamics, even as they impinge upon the acceptance of objective knowledge. For this is the business of history and sociology, and in particular, the history of mathematics and the sociology of its knowledge.
>
> *(p. 63)*

That said, Ernest's philosophy of social constructivism begs us to ask these questions as we go about our critical understanding of mathematics. We have thus opened the door to our final two sections of this chapter: looking to the history of mathematics and the field of ethnomathematics. It is my hope that these sections will help you to more fully appreciate how mathematics is a social construct. As you read, consider how these contributions describe a social process by which mathematical knowledge has been created.

What does the history of math tell us?

History of mathematics is an important field that, on its own, helps us to address the question of what is mathematics. It will help us to think more deeply about the social process by which mathematical knowledge has developed. Reading these histories reveals to us several things, including first and foremost the diversity of culture and people involved in the histories of mathematics. Also revealed in these histories are the close relationship between developments in mathematics and developments in society.

To start, histories of mathematics suggest the importance of a handful of ancient societies that contributed to modern Eurocentric mathematics. These include Egyptian, Babylonian, Greek, Roman, Hindu, Chinese, Japanese, Korean, and Muslim people. All made contributions to a variety of branches of mathematics that were further developed in the modern period. These groups developed such topics as computation, number theory, geometry, algebra, and applied mathematics. The following are some examples of these contributions. Read these examples to both experience the diversity in influences on modern Eurocentric mathematics and further your explorations of the concepts within mathematics.

Early mathematical practices existing in India and China are now very typical practices across the globe. "Decimal notation and the symbols for numerals we use today originated in India and came to Europe through the Arabs (Cooke, 1997, p. 197)." Over time and across societies, the format for each symbol representing the digits has changed. That is, there have been many different symbols to represent 1, 2, 3, … In particular, then, the Indian influence was the practice of using a single symbol for ten digits and then using these to describe any number as a sequence of digits. This saves us from continuing to invent new symbols for the various numbers. For example, we could have: 1, 2, 3, 4, 5, 6, 7, 8, 9, &, #, @, where & is a symbol to mean 10, # means 11, and @ means 12. Or, we could have our number 10, which relies on the concept of place-value and strings together two symbols together. The number 427 indicates that there are 4 hundreds, 2 tens, and 7 units. We can thus describe any number with only ten symbols and an understanding of place-value notation. The place-value practice in base ten was used by those in both India and China at the time. It has been difficult to determine who used it first, if that might be your interest, but

> It certainly came to the West from the Arabs, who learned it from India. In fact, one of the influential treatises by which Europeans learned about the decimal system and the symbols for digits was a treatise by the Muslim scholar Kushyar ibn Labban.
>
> *(Cooke, 1997, p. 197)*

Our numeral system is often referred to as Arabic, but has also been referred to by other names to reflect more accurate historical understandings, such as Hindu-Arabic Numerals.

Indian influence also included their dealings with number theory, or the branch of mathematics that studies whole numbers and rational numbers. Typical problems in the field include finding prime numbers and divisibility. For one, Hindus were interested in the triples of integers for which the sum of the squares of the two smaller equals the sum of the square of the larger. Two examples of these triples are the numbers (3, 4, 5) and (9, 40, 41). You may have encountered these before, under the name "Pythagorean triples." While the Pythagoreans may have been interested in their practical use as related to right triangles, it is possible the Hindus found a religious purpose to this project:

> A Hindu home was required to have three fires burning at three different altars. The three altars were to be of different shapes, but all three were to have the same area. These conditions led to certain 'Diophantine' problems, a particular case of which is the generation of Pythagorean triples, so as to make one square integer equal to the sum of two others.
>
> *(Cooke, 1997, p. 198)*

This example shows how the context within which mathematical knowledge originates can be surprising. Perhaps we might expect to have found the first use of these triples in a topic more relevant to engineering. The religious nature to this origination serves as an example of the socially constructed nature of mathematics.

Similar to number theory, algebra emerged among a variety of locations and cultures. Its title comes from the Arabic word al-jabr, used by Muhammad ibn

Musa Al-Khwarizmi of the ninth century. His work centers on solving equations with an unknown by keeping the equation balanced. This can relate to common practices in mathematics classrooms and modern algebra. For example, to solve the equation $3x + 9 = 12$, we can first subtract 9 from both sides to keep the equation balanced. In this way, Al-Khwarizmi was interested in developing an algorithm, or procedure, for solving equations with unknowns. This goal came about as the result of extensive work in dealing with equations with such unknowns. Many consider this as the essential feature of elementary algebra: solving equations to find an unknown value.

These examples from the history of mathematics aim to decenter a myth that modern mathematics is a Eurocentric conception. If anything, the pattern seems to be the appropriation of concepts originating in Middle East and Asian civilizations by European civilizations. Furthermore, the development of mathematics is rich with social contexts, and such histories help us to consider mathematics as a construction by groups embedded in social life. Other histories of mathematics put forth striking examples of the codevelopment of mathematics and society. Harouni (2015) demonstrates the close relationship between the school mathematics taught over time, emphasizing what he calls the type of mathematics that is useful for its "commercial-administrative" purpose. In this way, he demonstrates that the teaching of mathematics stressed primarily commercial-administrative mathematics, or those topics most useful for market economies. Think about all the mathematical exercises in schools that involves counting, money, buying and selling, "breaking even," and the like.

Similarly, Høyrup (1994) shares several examples of the codevelopment of mathematics with societal needs. In his discussions of Mesopotamian Mathematics, he refers to its role as a "bureaucratic tool" and the emergence of "scribal professional authority." The latter describes the emergence of a new skilled class of people in eighteenth century BCE who were valued because they demonstrated prowess and efficiency in computation and record-keeping.

Later in Høyrup's historical accounts of mathematical development, he calls us to focus on the entanglement between mathematical knowledge and state activities, including notably the crucial role mathematical knowledge plays in modern warfare. Historians of mathematics provide us with a more nuanced perspective in mathematics. Far from a white, male, Eurocentric project, mathematics over time draws from many cultures. Much of its development is clearly entangled with state developments.

Ethnomathematics: Thoughtfully considering an anthropology of mathematical knowledge

Having reviewed some examples from the history of mathematics, we move to the anthropologies of mathematics that further expand our conception of mathematics. This work, often referred to as ethnomathematics, suggests several important points. First, it reminds us that the roots of mathematical knowledge are computation, arithmetic, and geometry. It also suggests a novel answer to our main question that echoes Ernest's social constructivism: mathematics is *a language*. Finally, "ethnomathematics" continues to reject mathematics as a uniquely Eurocentric project, as the history examples above have done.

To begin, we should review just what is meant by the term. A major figure in ethnomathematics is Ubiritan D'Ambrosio. His (2002) survey of the field proves an excellent source to grasp its orientations and dispositions. As he puts it:

> Ethnomathematics is the mathematics practiced by cultural groups, such as urban and rural communities, groups of workers, professional classes, children in a given age group, indigenous societies, and so many other groups that are identified by the objectives and traditions common to these groups.
>
> *(p. 1)*

This claims that mathematics exists in a multiplicity of practices and, a point relevant to our teaching, that students and communities have mathematics embedded in their lives. Thus, D'Ambrosio suggests that ethnomathematics is essential to best practices in pedagogy.

One mathematics education scholar, Alexander Pais (2011), cautions us to think through more carefully about these applications of ethnomathematics. He provides examples when students have participated in a lesson in ethnomathematics, only to walk away with an unchallenged viewpoint of social relations. For them, the mathematical practices of "Other" cultures and people continue to be objectified. Pais also encourages the work of ethnomathematicians to focus more on the cultural practices of academic mathematicians themselves. This fits within the stated objectives for ethnomathematics laid out by D'Ambrosio's, but as Pais points out, all-too-often the field focuses more on seeking out the mathematics of the "Other" rather than understanding more fully how mathematical knowledge comes to be social constructed by academic mathematics. All in all, these efforts can provide more depth to the philosophical points made by Ernest regarding mathematics as a social construction.

Nevertheless, the contributions from ethnomathematics decenter our fixed conception of mathematics and deserve our attention. Several examples are contained in the edited volume *Mathematics across cultures: The history of nonwestern mathematics*. One author in this volume characterizes the field of ethnomathematics as having two branches: the first, a "general anthropology of mathematical thought and practice" in every geographic area, and, the second, the specific dedication to understanding the mathematical practices of small scale and indigenous cultures (English, 2000, p. 13). We have already uncovered much of the work in the former. For example, historical and anthropological inquiries have helped us to understand how the most widely used numeral system came from

FIGURE 1.4 The quipu from Incan civilization, knotting retains mathematical records

outside Europe. In other words, the first definition of ethnomathematics is in looking for these influences on modern Eurocentric mathematics. In contrast, the second definition does not concern itself with questions of influence and instead is determined to validate other cultures' use of mathematics. This project addresses the implication that mathematical behavior exists only in particular societies, such as those with written language, urban centers, agriculture, and/or hierarchical state structures.

Here are some examples of the findings in this second branch of ethnomathematics. The Incan use of the quipu has been well documented. These knotted ropes (see Figure 1.4) maintained records of transactions and tax collection and their patterns indicate significant mathematical computation beyond simple arithmetic. Mayan culture engaged significantly with mathematics: some paintings on pottery illustrate their math classrooms; they also had an understanding of zero, which you may not realize to be a major development in mathematical thinking.

Mathematical practices are also embedded in cultural practices, be they artistic, religious, or practical. People living in the Great Plains of Pre-Columbian North America constructed tipis, an architectural structure with several mathematical properties. Similarly, African American quilts that preserve personal histories via the use of colors, beads, and knots all have significant mathematics underlying them. Some might suggest we proceed with caution as we continue with such examples. Describing the mathematics of the tipi cone uses Eurocentric understandings and commits to the anthropological "gaze." That is, we are searching for the Eurocentric things embedded in indigenous cultures. In some ways, this is a project that elevates these cultural practices, because, previous to such engagement, they were deemed as inferior. On the other hand, we need to commit to the reality that any mathematical practice stands on its own as significant, not because it somehow resonates with a Eurocentric mathematical concept or practice. By looking to cultural practices and their embedded mathematics, we might add these practices to our set of mathematical activity, complementing mathematical operations and the various fields within mathematics. For example, quiltmaking is a mathematical activity.

Ethnomathematics provides other contributions to our understanding of the nature of mathematics, such as the suggestion that math is a language:

> Mathematics is a method for communicating ideas between people about concepts such as numbers, space and time. In any culture there is a common, structured system for such communication, whether it be in unwritten or written forms. These systems can form bridges of communication across culture and across time ... Mathematics has been part of all societies, a part of every profession as well as everyday life. Western mathematics became narrower with the insistence that only deductive mathematics from a set of axioms, following the Greek tradition, was *real* mathematics... [And,] mathematics has often worked on many levels, as part of everyday culture and also as used by subgroups within the main culture...Indeed in many cultures, the mathematics of calendars and astronomy were in the hands of the priestly classes.
>
> *(Wood, 2000, pp. 1–3)*

We see mathematical communication among all cultures and, interestingly, several cultures aimed to preserve some mathematical knowledge for a subset of its population. By viewing mathematics as a language, we also realize that mathematics exists anywhere there is communication, including cultures without written language. Mathematical communication does not have to be written down as evidenced by the intricate counting systems of Papua New Guinea and Oceania. Other examples are the weaving patterns of Northern Australian aboriginals and the knotting of quipus as discussed above. As we begin to shift our attention toward education, we can look at how communications can be mathematical. Communications include written, oral, and visual, through artwork or other representations, for example.

With our attention to ethnomathematics in this section, we have come to understand the major contributions as well as a caution to its pitfalls. It is important to remember that through ethnomathematics we come to fully appreciate what can be considered mathematical behavior. However, we need to think carefully about whether such thinking further objectifies "Other" cultures by studying them through Eurocentric eyes and, importantly as we transition to discussions of mathematics education, how we bring these discussions into the classroom.

In summary of this chapter, we have covered quite a bit of territory that can help us to understand what mathematics is. The introduction of mathematical operations and academic mathematics was a natural place to start, given that this is likely what we all think of mathematics in the first place. However, we next moved into some of the contests within the philosophy of mathematics and have put to you the question of whether mathematics exists without humans or was in fact constructed by them. A more critical viewpoint embraces Ernest's "social constructivism." Here, mathematical knowledge is the product of the mathematical community, itself embedded within historical and social contexts. These considerations from the histories of mathematics and the field of ethnomathematics decentered mathematics as an exclusively Eurocentric project. They also (along with the philosophy of mathematics exploration) emphasized the linguistic nature to mathematics, in which mathematical ideas are primarily communications between people. All of these considerations are highly relevant to teaching mathematics with a critical perspective.

At the conclusion of each chapter in this book, I provide activities and prompts for your consideration that expand on the contents discussed as well as classroom tips and a table to review the main concepts covered. You are encouraged to work collaboratively on these as well as take a look at the suggested readings from which I sketched this review of mathematics. For readers excited by these ideas, I suggest moving next to Greer's (in press) nuanced and advanced approach as he describes the contributions from Hacking, Høyrup, Hauroni, and several others I included in this chapter.

Activities and prompts for discussion:

1. Mathematics is typically viewed as an objective, value-free knowledge. For example, it is hard to dispute that 2 + 2 equals 4, correct? From your own experiences in mathematics and this chapter's sketch, how do you respond to the notion that mathematical practice is "black and white?" Do you side with a more absolutist or social constructivist philosophy of mathematics?

2. Browse the website http://www.storyofmathematics.com/ which presents a history of mathematics. Does this history as presented perpetuate a Eurocentric view of the history of mathematics? Does the website seem to align with a particular nature or philosophy of mathematics, such as those described in this chapter?

Classroom tips:

• For every mathematical topic you teach, attempt to learn histories of its development and include at least some focus on teaching with a historical lens. Whenever possible, focus on representing the multiple cultures involved, especially those that are not from European contexts.

• Break your students free from the notion that mathematics is simple a plug-and-chug, rule-following discipline. Integrate reasoning and discovery of mathematical ideas into your teaching so that you help students to see the joy and beauty of mathematical practice just as those throughout history discovered mathematical ideas for a multitude of goals and within varying contexts.

TABLE 1.1 Important terms and concepts in this chapter

Pure mathematics	Branch of mathematics focusing on proof and reasoning; a false dichotomy when relating to applied mathematics
Applied mathematics	Branch of mathematics applying formal mathematics to engage real-world problems; more accurately, pure and applied mathematics work hand in hand, such as with examples from mathematical modeling
Standard algorithm	The typical teaching approach to math skills that attempts to increase student efficiency but lacks opportunities for learners to appreciate reasoning and sense making when engaging mathematical concepts
Ethnomathematics	The anthropology of mathematical practices; studies *the people* doing mathematics, including practices embedded in cultural ways of being as well as the practices of academic mathematicians
Eurocentrism	A false notion that mathematics (and other disciplines) are developed mostly or entirely by European civilizations; teachers can trouble this notion through their teaching practices by including multiple histories of mathematical development
Codevelopment of math and society	Mathematical ideas developed hand in hand with developments in society, including motivations for spiritual understanding, practical solutions to social life (like building cities and large-scale agriculture), the development of market economies, and the needs of emerging state governments; at present, many mathematical ideas emerge for the needs of a military-industrial complex of state governments
Platonism and its offshoots	In these philosophies of mathematics, mathematics is a universal truth that humans discover over time; entangled with a religious quest committed to unlocking the mystery of a higher power
Fallibilism and social constructivism	Other philosophies of mathematics declare it as a social activity and reject the notion that mathematics is a universal truth; instead, mathematical communities accept new ideas because they are not yet proven false

Further reading

Cooke, R. (1997). *The history of mathematics: A brief course.* Wiley.

D'Ambrosio, U. (2002). *Ethnomathematics.* Sense.

English, R. (2000). Anthropological perspectives on ethnomathematics. In H. Selin (ed.) *Mathematics across cultures: The histories of non-western mathematics*, pp. 13–22. Springer.

Ernest, P. (1991). *The philosophy of mathematics education.* Routledge.

Greer, B. (in preparation). Why/how people develop mathematics. In B. Greer, O. Skovsmose, & D. Kollosche (in preparation) *Breaking images: Iconoclastic analyses of mathematics and its education*. A volume in the series "Studies on Mathematics Education and Society" Open Book Publishers.

Hacking, I. (2014). *Why is there philosophy of mathematics at all?* Cambridge University Press.

Harouni, H. (2015). Toward a political economy of mathematics education. *Harvard Education Review 85*(1): 50–74.

Hersh, R. (1997). *What is mathematics, really?* Oxford University Press.

Høyrup, J. (1994). *In measure, number, and weight: Studies in mathematics and culture*. SUNY Press.

Lockhart, P. (2009). *A mathematician's lament: How school cheats us out of our most fascinating and imaginative art form*. Bellevue Literary Press.

Pais, A. (2011). Criticisms and contradictions of ethnomathematics. *Educational Studies in Mathematics 76*: 209–230.

Selin, H. (Ed.) (2000). *Mathematics across cultures: The history of non-western mathematics*. Springer.

Wood, L. (2000). Communicating mathematics across culture and time. In H. Selin (Ed.) *Mathematics across cultures: The histories of non-western mathematics*, pp. 1–12. Springer.

2

REFORM MATHEMATICS TEACHING

The student-centered approach

With the preceding exploration of mathematics at hand, we now turn to our work that occupies the remainder of this book: critically examining mathematics education. In this chapter, we'll focus on a review of what the mathematics education research community has prioritized since the 1960s, often referred to as a reform mathematics education that has emerged from new understandings in educational psychology. Mathematics education researchers have drawn from sociocultural and constructivist learning theories to emphasize student-centered teaching approaches over the traditional skills-based instructional model. There has been some backlash to these approaches, dubbed the "math wars," but ultimately the research-based practices from reform mathematics research require classroom mathematics teachers to thoughtfully consider every classroom lesson as a social experience from which each student makes meaning for mathematical content. The chapter provides clear guidelines on how to develop mathematics lesson and unit plans that provide enough student-centered experiential and social learning that also meets the breadth of typical curricular goals in mathematics education standards.

Reform mathematics teaching: Pedagogical content knowledge and student-centered learning

Reform mathematics teaching has prioritized the importance of mathematics teachers' knowledge and practice of specific pedagogies that increase student understanding of the content. Mathematical pedagogical content knowledge is a term to describe these specific practices and we will look at prominent researchers in mathematics education who provide important ideas and motivations for mathematics teachers to think differently about how they approach classroom

DOI: 10.4324/9781003322566-2

instruction. By considering their research, we can take mathematics instruction away from a scripted experience of practicing mathematical skills and toward a space where students appreciate mathematics more broadly.

Think of pedagogical content knowledge as the intersection of your knowledge of mathematics and your knowledge of teaching. For example, you may be highly proficient in mathematics, able to compute mental mathematics efficiently, or to carry out a complicated derivative in a calculus exercise. As well, you also may have been taught important ideas about how to organize a classroom, how to design an engaging lesson, and how to deliver a lesson effectively. However, this is not enough, you need to have a robust understanding of the specific pedagogies and methods for *teaching mathematics*. Lee Shulman, a celebrated education researcher not specific to mathematics teaching, coined the term "pedagogical content knowledge" as the intersection of the content knowledge that you have and the teaching practices that you do. In other words, the specific practices that you employ in a mathematics classroom, the specific areas that you plan for and attend to during delivery, will be slightly different than that of someone teaching an English or social studies lesson.

For mathematics pedagogical content knowledge, Ball, Lubienski, and Mewborn (2001) point our attention to the importance of mathematics teachers' understanding of mathematics as the cornerstone of pedagogical content knowledge. They open with the following "tests" of our own understandings of mathematical knowledge. How many of these questions are you able to answer?

> Why does it work to add a zero on the right when multiplying by 10, or two zeros when multiplying by 100? Why, when the number includes a decimal, do we move the decimal point over instead of adding zeros? Is zero a number? If it is a number, is it even or odd? What does it mean to divide *by* one-half? What is an irrational number? Is a square a rectangle? What is the probability that in a class of 25, two people will share the same birthday?
>
> *(Ball et al., p. 433)*

Many of the examples are focused in elementary content knowledge and Ball and other's research point to the reality that several elementary teachers of mathematics might be stumped by these questions. They provide an important answer to the failures of traditional mathematics education: we focus our classrooms on practicing mathematics without the understanding behind it. True, some students will be able to follow the procedures taught to them without an understanding of the content, but the research shows that for several students this will not work. Mathematics teaching needs to provide students the space to ask questions and see the mathematics for themselves.

As these ideas developed in the research, mathematics education policies prioritized teaching both the process of mathematics and the content. The National Council of Teachers of Mathematics (NCTM) provided their blueprint for a

TABLE 2.1 Common core's standards for mathematical practice

1. Make sense of problems and persevere in solving them.	2. Reason abstractly and quantitatively.	3. Construct viable arguments and critique the reasoning of others.	4. Model with mathematics.
5. Use appropriate tools strategically.	6. Attend to precision.	7. Look for and make use of structure.	8. Look for and express regularity in repeated reasoning.

national math curriculum in the United States first in 1989. You'll see an emphasis on teaching students the why's of mathematics as well as the how-to's of mathematics. These standards were revised in 2001 and continued with the twin emphasis on process and content. In the United States, most state standards are now informed by the 2010 Common Core mathematics standards which, again, continue to emphasize both mathematical content and process. The Common Core's eight "Standards for Mathematical Practice" are included in Table 2.1.

The Common Core standards were clear to also express detail in the "how-to" of mathematics, but notice the emphasis on student work that requires them to think critically about the mathematics. Put another way, emphasizing the process of mathematics in the classroom means providing opportunities for students to behave like "mini-mathematicians." Remember the discussion of pure and applied mathematicians in the previous chapter? Our students should be asked to struggle with mathematical ideas together, at their levels, similar to how pure and applied mathematicians negotiate and understand new ideas. The first step in doing this is for a mathematics teacher to make sure to check their own deep understandings of the mathematical content. Next, they'll be able to design activities and experiences leading students to act in these ways.

When I discuss these ideas with mathematics teachers who have only experienced traditional math teaching, they often ask me, "What in the world does this look like?" Fortunately, we have several great examples from the research including several demonstration videos of teachers working with students in public school classrooms. Another celebrated math education research is Jo Boaler. As a university researcher, she coauthored a book with demonstration videos alongside her teaching colleague Cathy Humphreys, *Connecting mathematical ideas: Middle school video cases to support teaching and learning* (2005). As one example, Humphreys leads her classroom of students on a 30-minute exploration of the answer to the question: "What is 1 divided by two-thirds?" As the students begin in earnest, she quickly clarifies that some students may have learned "a rule" at some point but that she is not interested in the answer determined using the rule; she wants students to reason what the answer is and why it makes sense.

Another excellent example of reform mathematics teaching videos comes with the website from the Dana Center in Texas. Uri Treisman, another major figure in reform mathematics teaching, founded the center and promotes best practices in mathematics instruction via professional development for mathematics teachers and dissemination of best practices through the center's website. The site contains several videos spanning mathematical topics across content and grade level, and in each of these you can witness the ways that a teacher's deeper understanding of the content allows for students to engage in the process of mathematics.

I encourage you to watch several videos of reform mathematics teaching, including those suggested above and also live in person with teachers you know who practice this already. In the next section, I provide specific guidelines on how to approach the design and enactment of teaching reform mathematics lessons. As the research reviewed shows, the first and most important step is to ask yourself: "What is the deeper meaning in the content that I am about to teach? Is there a big idea or concept?" From there you will be able to think about the conceptual understandings or experiences that your students are bringing to the lesson and how you will set the stage for an experience that draws from these and provides experiences in process-based mathematics.

Critically planning mathematics lessons

In the previous section, we reviewed the goal of reform mathematics teaching that emphasizes the process of mathematics. Here we will look at how to go about designing classroom experiences that provide this for your students. Although reform mathematics teaching prioritizes these types of lessons, it does not suggest we entirely remove mathematical practice and skills development. Reform mathematics teaching stresses that we provide opportunities to learn the process of mathematics and to see the "big ideas" at play, but it also stresses that students must be proficient in mathematical skill and practice. For this reason, we will also review the best lesson planning methods for helping students to practice skills with efficiency. Ultimately, as you design mathematics lessons, reform mathematics teaching asks you to typically start with the conceptual understanding with a few days of experiences in process-based mathematics and then move to skills development and proficiency. We also want to make sure our assessments of student learning prioritize both the concepts and skills relevant to the topic at hand.

When I start a new mathematics unit, I typically open with an experience-based lesson that gives students the opportunity for some process-based mathematics. I think carefully about the big concepts at play in the new unit and design an experience that will get students to "see it" first hand. Sometimes this may be a lesson that spans a few days or even a full week. Table 2.2 includes three different examples of opening, experience-based lessons for some units across the

TABLE 2.2 Experience-driven first lessons for three different mathematics units

Unit topic: Two digit addition	Grade level: 2nd	Prior conceptual knowledge: –Place value –Addition

Opening experience: In small groups, students will be given "story problems" that require solving two-digit addition problems together. They have not been instructed how to do these before and are encouraged by the teacher to think about their understanding of place value and the concept of addition to "invent strategies" to answer the question. Ultimately, students will discover mental strategies that, over time, they can use for two-digit addition problems and help them to understand the standard algorithm much better.

Unit topic: Systems of equations	Class: Algebra 1 or 2	Prior conceptual knowledge: –4 representations of a function (table, equation, graph, real-world context)

Opening experience: In small groups, students are given two scenarios of taxi-cab fares, one for city A and one for city B. They are asked to compare the two fares and think about how one might be better or worse, or how this might change based on the distance traveled in the ride. Ultimately, students will discover the idea of a solution to the system and see it represented in the equation, table, and graph.

Unit topic: Polar coordinate functions	Class: Precalculus or Trigonometry	Prior conceptual knowledge: –Cartesian coordinate plane –unit circle angle measurements

Opening experience:
In partners, students are provided two axes with a single point marked in one of the four quadrants. They are also given rulers and protractors. They are tasked to provide complete information for another pair of students to locate the exact point. Students usually describe the horizontal and vertical distance from the origin quickly, and next are prompted to find another way. After some time and encouragement to use the protractor, they realize that they can also describe the point as its distance from the origin plus the angle it makes in standard position as with the unit circle.

curriculum. I also provide the prior conceptual knowledge that I draw out from my students as they engage in the process.

In each of the three examples, the goal is to provide a rich experience for students to engage and debate together. Ultimately, the students will reason and explain big ideas for the unit at hand together in small groups and in whole-class discussion. What has been experienced and understood together is a big idea that will continue throughout the whole unit, something that we can continually refer back to as we get into the intricacies of skills and practices for the topic. For the second graders, we will continue to think about place value and our "invented strategies" as we practice two-digit addition. For the systems of equation unit, we will have a rich example of what a "solution to a system" means and will understand it as the ordered pair that satisfies both equations, as the intersection of the two graphs, etc. For the polar coordinates unit, because they "invented" (r, theta) as the way to locate a point on a graph, they will be much better at using polar coordinates as they work with functions using them.

The first step in planning these experiences is thinking deeply about the concepts at play and what your students bring to it via their previous conceptual knowledge. The next step is designing/selecting the task and structuring it in a way that will lead to maximize productivity as they work through the mathematical process. I like to think of the experienced-based lesson with a format to include the following components: hook, task in small groups, whole-class discussion, and closure. This format is informed by reform mathematics teaching research, especially in how to design a task and how to scaffold the experience so students can get the most out of it.

Starting with the hook, as with all lessons, you want to get students engaged with the work at hand and have them all on the same wavelength together. This can often require some "non-math" talk about some of the context to the story problems you will have students working on in the lesson. For the algebra systems lesson, you can start with a whole-class discussion about student experiences or knowledge with taxi-cab rides. "Does anyone have a family member who drives a taxi? Has anyone ridden in a taxi? How does taxi fare get calculated?" Structure these conversations with turn and talks to get everyone's mind on the topic. Or, if it works for your students, your hook can approach some of the relevant math right away. You can include some discussion of the prior conceptual knowledge that you'll want students to have at the forefront of their minds. For our second graders, some discussion about place value might be a good approach. "Can you and your partner arrange the 23 beads on your desk in a creative way? Think about grouping them as we have done before."

After the hook, the teacher needs to set the stage for the experience clearly for everyone to understand what they are being asked to do. The task selection is important as well as the role of collaboration with peers. The teacher "guides and facilitates, poses challenging questions, and helps students share knowledge" and the students "work in a group and learn actively" (Kuper & Kimani, 2013). Many have misunderstood this to mean that teachers are "told not to tell," implying that teachers take an inactive role. However, this is entirely not the case; teachers must take an active role in selecting the task that will lead to learning, in anticipating how students will approach and work through the task and how they will probe and press students toward learning, and in orchestrating the subsequent whole class discussion. For the latter, Chazin and Ball (1999) give an especially thorough account of the active role mathematics teachers must take to maintain the discussion's productivity and goal.

The central experience, or what many call the task, makes or breaks the experience-based mathematics lesson. Stein and Smith (1998) have codified the language of task selection into lower- and higher-level cognitive demand. The examples contained in their article provide descriptions of what kinds of tasks push toward the mathematics lesson that engenders significant, deep meaning-making. As they describe it, a task

> is defined as a segment of classroom activity that is devoted to the development of a particular mathematical idea. A task can involve several related problems or extended work, up to an entire class period, on a single complex problem. Defined in this way, most tasks are from twenty to thirty minutes long.
>
> *(p. 269)*

The higher cognitive demand categories are "procedures with connections" in which students make meaning as they recall and connect procedural mathematics and "doing mathematics," where students uncover answers to previously unknown questions.

Equally important to the selection of tasks is the orchestration of mathematical discussions that will unpack the experience of engaging with the task. This phase of the experience-based mathematics lesson has been studied significantly by researchers as well, such as in the article by Smith, Hughes, Engle, and Stein (2009). Their research argues five practices to successful unpacking of experiences in the whole-class discussion:

1. Anticipating student responses to challenging mathematical tasks;
2. Monitoring students' work on and engagement with tasks;
3. Selecting particular students to present their mathematical work;
4. Sequencing the student responses that will be displayed in a specific order; and
5. Connecting different students' responses and connecting responses to key mathematical ideas.

(p. 550)

This helpful list promotes more productive struggle for students. The idea is not to turn students over to a task that is out of their reach, but to support their sensemaking throughout and especially when the whole class debriefs the experience after working on it in small groups. Now that you have this structure in mind, you might want to rewatch the videos mentioned earlier and notice the teacher moves made that correspond to these strategies.

Although advanced reform math teachers incorporate these kinds of experienced lessons often into their instruction, a beginner who is attempting to shift instruction away from the traditional model can set a goal to start every unit in this way. Having a rich experience to develop the concept and big ideas in the unit will provide meaningful connections for the work throughout the unit. As students move into the skills and practices needed for the topic at hand, teachers continue to refer to the big ideas developed at the start of the unit. How we structure the more skills-oriented knowledge for the unit is also important to consider and we turn to this lesson plan format next.

A direct instruction lesson for mathematics will sound like the traditional approach but a reform-minded teacher knows that this must remain student centered and very active. The lesson begins similarly, with a hook to generate interest and/or activate prior knowledge. The next segment of the lesson after the hook is the modeling phase. To be effective, such a direct instruction style lesson must incorporate active learning methods informed by cognitive research that tells us how people learn. In modeling, a teacher will explain the definitions, concepts and skills directly to the students but will do so with an active student-centered approach. This takes on a variety of forms as well and depends on the learning goals. If the lesson is teaching a procedural skill, the notion of a think-aloud is highly appropriate. A think-aloud is when the teacher models the problem by noting every detail in their thought process as they complete it. They articulate out loud all the information they receive, all the choices they have, and which choice they make and why. Here is an example for thinking aloud how to graph an equation using a table of values, with actions and gestures the teacher takes put in italics. The equation to graph is $y = x^2 + 1$.

> I need to graph the equation $y = x^2 + 1$. *Points to equation.* Because the equation has only x's and y's and I need to graph it, this means I will need an x-y plane. *Draws a coordinate plane on the board.* Now, I need to think about what this will look like. *Pause.* I know that sometimes when I graph equations they turn out to be lines and other times they turn out to be curves. *Pause.* I'm not quite sure what this will look like because I haven't done enough of these to know a pattern yet. But I do know that graphing means plotting the points that "work" *(uses air quotes)* for the equation. I mean, I know that I need to find numbers that I can put in for the letters in the equation and then I can plot these on the graph and try to find a pattern. Let me see if I can try to find a pair that "works." So, how about putting a 1 in for x? *Takes a notecard with the number one on it and places it over the x in the equation.* So, putting a 1 in for x means that I need to square it and add 1. But which do I do first? Let me think. *Pause.* Order of operations tells me that exponents come before addition, so I square 1 first to get 1 and then add 1 to get 2. *Records this on the board.* This means that when x equals 1, y equals 2, and I think I need a way to record this better as I work with more numbers. *Pause.* Oh yes, I remember, I can use a table of values to record the pairs that "work" for the equation.

The think aloud for the problem continues similarly. Envision a teacher performing this think aloud slowly, clearly, and with engagement with her audience so that the students are invested in the outcome. It is almost like a story being told.

There are several other methods to make the modeling portion active for learners. A teacher can question the students on what to do next by calling on students or stopping briefly for a think-pair-share. The latter is when the teacher poses a question to the class, requires students to think individually for a designated moment of time, next share their thoughts together for a designated amount of time, and, finally, pairs are called on to share their thinking. Think-pair-shares are highly effective for use in any part of a mathematics lesson and for either direct or indirect instruction lessons. A final way to keep students active in the modeling phase is the use of guided notes; however, these should prompt for students to thoughtfully respond on the page rather than fill in blanks or copy work from the board.

These possible methods for increasing active learning during the modeling phase are simple examples to move mathematics teaching in the reform direction. As this book takes on the topic of race, class, gender, sexuality, language, and ability in future chapters, you will want to revisit the notion of increasing access when explaining mathematics to learners. For example, some believe that all mathematics instruction requires mathematically precise language. However, notice how in the think-aloud example above, the language used did not use precise mathematical language with well-defined terms throughout. While mathematical accuracy and precision is an important aspect of mathematical practice, a critical perspective on teaching mathematics recognizes that mathematical communication can take a variety of forms that induct learners *over time* into the practice of refined mathematical communication.

After the modeling section, a direct instruction lesson requires students to apply the information they received immediately. This happens in two phases: guided and independent practice. There are several ways to structure this, and effective approaches emphasize quick feedback to learners and social engagement with other learners. In guided practice, a teacher can pose a problem to the class and have students work in partners to solve together. This can happen for several rounds before students work independently on a set of problems for individualized practice. Social learning is particularly important in both phases, as peers working together can negotiate meaning faster than in isolation. There are lots of ways to structure guided practice that will keep students active and engaged, and this depends greatly on the community that the teacher and students establish.

Both indirect and direct lesson plans end with a closure. Like the hook, closures contain several objectives and you should strive for your lessons to close with as many of these aspects. First, a closure needs to provide an opportunity for the student to walk away with a clear summary of what was learned during the lesson. This can be as simple as a teacher restating the main content objective one last time, but it is best to first have the class state this and teacher refine if necessary. As well, a closure can provide the teacher with a check for understanding. Finally, the closure connects back to the opening and relays what the next steps will be, either in how the material learned will be used independently or in work together in the next lesson.

To summarize, a beginning reform mathematics teacher can set a goal to start each unit with the big ideas and concepts at the core of the mathematical topic and start their students off with an experience-based lesson to "do mathematics" and engage in the mathematical processes of reasoning and sensemaking. The first step here is determining the mathematical task, and there are several places to start including some of the video resources mentioned above. Next, the unit can proceed with active student-centered lessons that use the structure of modeling and guided practice but continually reinforce the big idea of the mathematics at hand. Consider the following activities, classroom tips, table of important terms and concepts, and suggested further materials as you continue to learn more about reform mathematics teaching.

Activities and prompts for discussion:

1. Would you say that your mathematics teachers in the past have favored traditional or reform teaching methods? What style do you seem more comfortable with and which seems harder?
2. Pick a mathematical unit for a grade level that you might be asked to teach. Think deeply about the big idea in the unit, using resources to help you when necessary, as you begin to develop a first lesson that is experience driven. Sometimes your assigned textbook to use with students will have an "explorations" activity in most chapters. These are often too directed but can provide inspiration on how you could design an experience to start the unit. Remember that the experience should give a strong foundation on the big conceptual idea in the unit.

Classroom tips:

- Open every new unit with an experience rich in the mathematical process. Make sure students have the opportunity to investigate and inquire about new mathematical ideas before you teach them anything with a more direct approach.
- When practicing skills, make sure to keep these approaches very student-centered and with ample opportunity for students to work together as they practice.

TABLE 2.3 Important terms and concepts in this chapter

Reform mathematics teaching	Since the 1960s, research in math teaching shows that highly effective math teaching must include teaching students the process of mathematics in addition to math skills; our classroom instruction must include experiences and tasks to understand main concepts in mathematics

(Continued)

TABLE 2.3 (Continued)

Math pedagogical content knowledge	The intersection between a teacher's knowledge of math and knowledge of teaching practices; mathematics teachers need to understand the deeper concepts at play in every mathematical topic they teach
Content vs. process-based mathematics	Mathematical content are the typically thought of skills and concepts that are taught in the curriculum; process-based mathematics are the mathematical processes we teach in the curriculum like problem-solving, reasoning, proof, and mathematical modeling
Mathematical task	The central activity of a reform mathematics lesson; working in small groups, students collaborate and think together about something that they have not worked on before or been taught to do yet; teacher plays an active role by prompting students to recall relevant prior knowledge and develop their reasoning as they work toward a solution
Modeling	In a direct instruction lesson, the teacher clearly models the steps toward a solution, doing so by making clear connections to the conceptual ideas in the problem; a think aloud is a very important tool when modeling an example
Guided practice	After the modeling of a skill, students actively work together to practice with support and monitoring from teacher

Further reading

Ball, D.L., Lubienski, S., & Mewborn, D. (2001). Research on teaching mathematics: The unsolved problem of teachers' mathematical knowledge. In V. Richardson (Ed.), *Handbook of research on teaching* 4th ed. Macmillan.

Boaler, J., & Humphreys, C. (2005). *Connecting mathematical ideas: Middle school video cases to support teaching and learning.* Heinemann, 433–456.

Chazin, D., & Ball, D. (1999). Beyond being told not to tell. *For the Learning of Mathematics* 19(2): 2–10.

Dana Center. (2022). Public lessons. https://www.insidemathematics.org/classroom-videos/public-lessons

Kuper, E., & Kimani, P. (2013). Responding to students' work on a rich task. *Mathematics Teaching in the Middle School* 19(3): 164–171.

Smith, M.S., Hughes, E.K., Engle, R.A., & Stein, M.K. (2009). Orchestrating discussions. *Mathematics Teaching in the Middle School* 14(9): 549–556.

Stein, M.K., & Smith, M.S. (1998). Mathematical tasks as a framework for reflection: From research to practice. *Mathematics Teaching in the Middle School* 3(4): 268–275.

3

WHY IDENTITY, HUMAN DIVERSITY, AND INTERSECTIONAL IDENTITIES MATTER TO MATHEMATICS EDUCATION

In this chapter, we move from our discussion on mathematics and reform math teaching to the book's primary focus on human diversity in the mathematics classroom. I introduce relevant ideas from the social sciences starting with the broad concepts including facets of human diversity, social construction, and intersectionality. I move next to examples from performance data in learning outcomes in mathematics education that demonstrate that our current pedagogies are not effective for all learners. Even with the reform efforts from Chapter 2, all students are not provided equal opportunities to learn, enjoy, and apply mathematics. Finally, we will think about mathematics education's new focus on human diversity in the classroom, often referred to as its "sociopolitical turn" that reveals why identities matter in the mathematics classroom.

We hear the term "diversity" a lot in today's world so it is important to take a step back and think more carefully about what we mean by it. The human population contains a variety of identities to which individuals and groups align. On the one hand, these identities offer individuals and groups a sense of belonging and affirmation of their personhood and group experience; on the other, some identities receive significant discrimination in society based on preconceived notions of inferiority or "being different than the norm." You may have noticed institutions like a workplace or a school have listed what they call the "facets" or "aspects" of human diversity. These lists demonstrate the varying identities at play in human diversity and are considered for the ways that they both affirm a person's individuality, belonging to a specific culture or group, and make clear that individuals and groups can be subject to oppression when in the non-dominant group. We often say that those identities that are not in the dominant group are marginalized or sometimes referred to as the minority group. These days, we use the term "minority" not as a noun but rather as a qualifier for a

DOI: 10.4324/9781003322566-3

particular group of people. We'll discuss other language practices a bit more later in this section when we look at bias-free communication. For now, it's helpful to dig into the facets of human diversity by looking at the list of identities usually at play. A typical list of the facets of human diversity includes race/ethnicity, social class, gender, sexual orientation, ability, language, geography, religion, and marital/familial status. Notice that we cover most of these in future chapters of the book specifically as they relate to teaching mathematics.

Key concepts in diversity

Each facet of human diversity is highly complex and different from each other, and in the subsequent chapters we will explore each in greater detail. However, there are considerations that generally apply to all facets of human diversity. The first one we will approach is what social scientists call a "social construction." Each of the facets of human diversity has been determined socially by people and groups of people over time. For example, there is no biological or genetic basis for one's race or ethnicity. Similarly, gender is not biologically based. Assigned sex at birth is something that is biologically based, but as we will discuss in Chapter 6, gender is not. When we say that identities and facets of human diversity are social constructions, we are not saying that they are fake or that they do not matter. We are also not saying that one's identity is a choice. Rather, we are saying that these identities have come about socially and changed over history and that one's identity is important to their personal and social life.

For example, race is a social construction that places people into categories mostly based on their physical appearance including but not limited to their skin color. Given a history in the United States that results in white supremacy (to be covered in greater detail in Chapter 4), many people of color experience racial prejudice and discrimination. Unfortunately, many of the stereotypes and prejudicial thinking have been taught to us throughout society's various outlets, like school textbooks and the media, to the point that people will think prejudicially on a subconscious level. As a result, a teacher may have prejudicial thinking about a student of color in their class, thinking that they are not going to perform well academically and/or be more likely to misbehave during instruction. When a teacher does not confront these ideas carefully, they will continue to think prejudicially mostly because society has given them this message time and time again.

Thinking negatively about a person based on their identity is prejudice. Acting on this thinking is discrimination. Continuing our example, a teacher acts on their prejudice in one way by over-disciplining their student of color. For example, they might hold them strictly to the classroom rules, whereas they might "give a pass" to a white student when they make the same infraction. This often happens at the subconscious level and requires teachers to vigilantly confront the ways that society has messaged them about the various identities of students in their classroom. Eliminating prejudicial thinking is the final goal, but the first place to start is to recognize the prejudicial thinking when it happens and stop any discriminatory actions from taking place.

Newer concepts related to prejudicial thinking and discriminatory actions are implicit bias and microaggressions. Implicit bias suggests how society's messaging about marginalized groups causes individuals to have subconscious prejudices that, sometimes, also cause individuals to make discriminatory actions. At this point, the reality of implicit bias is hard to refute given projects like the implicit association tests run by social scientists at Harvard University. It is a useful exercise for teachers to take these tests to know more about their own implicit biases. You can point a web browser to their tests at https://implicit.harvard. edu/implicit/takeatest.html and choose from among tests that demonstrate your potential subconscious biases related to varying ethnicities, genders, sexualities, and religions.

One of the simplest but very damaging discriminating actions that someone can take as the result of their subconscious biases is called a microaggression. When acting on our biases we might make a subtle comment or take a simple action that would not be characterized as outright, blatant discrimination but is rooted in implicit biases and extremely harmful to the person receiving the action. You can find many examples of microaggressions as this is a popular topic being addressed by several outlets including academic institutions and workplaces. Racial and gender microaggressions are the most common examples to find, and you can imagine how they can happen for other marginalized identities as well. One type of microaggression is called a microinvalidation. As an example of a racial microinvalidation, consider this scenario: a white person listens to an experience of a Black person who has felt they were just racially profiled when shopping in a store. One of the store staff was following them around from a distance, and when they went to leave, a different store member asked to check their backpack, assuming shoplifting. When hearing this story, the white person replies with something like, "This was likely not because you are Black, maybe they were doing that to everyone today. Come on, that doesn't really happen anymore and I know that's a good store with good employees who are well intentioned." This reply invalidates the Black person's experience in the store, which is not an isolated event but likely a pattern they have experienced multiple times given the systemic racism that occurs throughout society. The white friend has committed a racial microaggression which further harms their Black friend who just experienced racial profiling while shopping in a store. Generally, as teachers, we should learn about different types of microaggressions, view several different examples, and think about how these might come up in the classroom.

These ideas of prejudicial thinking and discriminatory actions show us that identity causes significant problems through interpersonal relations and interactions. We should not think, however, that this is the only way that oppression comes about in society. In fact, many argue that the interpersonal interactions are just one part of the equation and that most of the power and oppression occurs at the societal level. Systemic or institutional oppression, such as systemic racism, refers to the ways that societal structures repeatedly disadvantage individuals in certain identity groups. There are several examples of this throughout history and in the present day. As one example, we continue to have wage and compensation differences between men and women who perform equivalent occupations (Barroso & Brown, 2021). As another, we have disproportionate rates of traffic stops by race in the United States (Baumgartner, Epp, & Shoub, 2018). Although traffic stops, as an example, technically come about through interpersonal interactions, it is the repeated pattern in the policing process that is the systemic racism. This takes us into a different approach about thinking about new policies and practices to confront systemic oppression for identity groups. We can think about standard practices in institutions and systems, like the government's police forces and our public schools, and make steps to address oppression at the institutional level.

Another important consideration when working with diversity is that each identity an individual has does not work in isolation but together with all identities present in their full makeup. The social sciences refer to intersectionality as the concept where identities interact at both the individual and group level. Vivian May refers to our typical approach as a "single axis" world where too often one of the social identities is privileged over another. Intersectionality promises to carry forth the traditions set up by third wave and Black feminists, a tradition that we'll look at again in Chapter 6's focus on gender. Black feminist author bell hooks has provided accounts and theorizations of feminists within both the civil rights movement and the feminist movement that fell short of describing and accounting for her Black female experience. Worse, in failing to acknowledge the multiplicity of oppression, both movements in their own ways served to reinforce oppressive structures for Black women.

May (2015) describes intersectionality as follows:

> First, it approaches lived identities as interlaced and systems of oppression as enmeshed and mutually reinforcing: one aspect of identity and/or form of inequality is not treated as separable or as superordinate. This 'matrix' worldview contests 'single-axis' forms of thinking about subjectivity and power and rejects hierarchies of identity or oppression. An intersectional justice orientation is thus wide in scope and inclusive: it repudiates additive notions of identity, assimilations models of civil rights, and one-dimensional views of power... For instance, its matrix model changes the terms of what 'counts' as a gender, race, sexuality, disability, nation, and/or class issue or framework. Intersectionality also approaches lived identities, and systemic patterns of asymmetrical life opportunities and harms, from their interstices, from the nodal points where they hinge or touch.
>
> *(p. 3)*

Thus, we should not prioritize racial equity or gender equity, for example, and not treat these as separate issues in isolation from one another. Instead,

> By focusing on how patterns and logics interact, and how systems of oppression interrelate, intersectionality highlights various ways in which, unwittingly, we can be engaged in upholding the very forms of coercion or domination we seek to dismantle. It is thus indispensable for identifying paradoxical outcomes and for revealing unexpected or hidden points of contact between the liberatory and the coercive. Via its matrix orientation, and attention to relational power and privilege, simultaneity, and underlying shared logics, intersectionality needs to be understood to have explanatory power, analytical capacity, and normative political component, one focused on eradicating inequality and exploitation.
>
> *(p. 5)*

Mathematics education scholarship has begun to reveal the natures of such intersections. Riegle-Crumb and Humphries (2012) researched how math teachers discuss their students' performance paying particular notice to how they talk about different gendered students. Their research revealed that these discussions also contained hidden racial connotations as well: "Stereotypes of men's innately higher math ability refer specifically to white males" (p. 295). Their research provides further study of the complicated nature of teacher perception on ability, in which the multiple social identities that learners occupy is reflected in a teacher's perception of the student's ability. These are the types of discussions that the lens of intersectionality affords and that mathematics educators can reflect on. Holding a single-axis view, as so many researchers and teachers have done, fails to acknowledge both the systemic nature to oppression across social identities and the multiple oppressive structures that learners will face within the school structure.

We've covered important topics related to human diversity including prejudice, discrimination, systemic oppression, and intersectionality. A final point that is important to discuss in this introductory description of working with human diversity is the importance of our language practices. The way that we speak about diversity and the identities within it is a matter of great significance. You may have heard about the term "political correctness" and have a reaction to it either positively or negatively. Social scientists and others who advance our thinking about human diversity have moved far beyond the notion of political correctness especially because the name does not capture the spirit of doing better with our language practice. Nowadays, you can find several different guidelines for a concept called "bias free communication" which better represents our goals as we strive toward strong language around human diversity. A common goal in these guides includes making sure to attend to the ways that a social group or an individual prefers to be identified. For example, as we will discuss in Chapter 8's focus on ability, some people with disabilities prefer "person first language" (e.g., a student with Autism), whereas others in the community prefer it the other way around to affirm their identity as a disabled person.

With its organization's focus on reducing oppression and discrimination against Jewish people, the Anti-Defamation League has a guideline for bias-free communication with several tenets that are listed in Table 3.1. For example, the third suggests that oftentimes we refer to someone's marginalized identity when it is not relevant. A simple example you might have seen before: a news article will state "twenty-two House Democrats, led by gay Democratic Representative Barney Frank," whereas you would not find "Michael Bloomberg, noted heterosexual mayor of New York" (ADL, 2014, n.p.). The sexual orientation for each politician was irrelevant in each statement so should not have been included; it is problematic because noting the minoritized sexual identity of Barney Frank, an irrelevant aspect to the political statement, reinforces his identity negatively. Read carefully these guidelines for bias-free communication (which were adapted from an older discussion of these practices) and think about the habits of language practice that you might have and whether they accord or conflict with the guidelines.

Although the guidelines are especially developed for writers, these are highly relevant and adaptable for teachers as well. For the last one, think about how many times our mathematics textbooks and posters fail to represent all identity groups. One teacher action to consider is a review of all materials and increasing the representation of identities, including women, people of color, and LGBTQ+ mathematicians into our classroom environments.

In beginning our work about human diversity as the starting point to explore its relationship to the mathematics classroom, we have reviewed the concepts of social construction, prejudice, discrimination, implicit bias, systemic oppression, intersectionality, and bias-free communication. Next we will look at the

TABLE 3.1 Tenets of bias-free communication, ADL

1. Be aware of words, images, and situations that suggest that all or most members of a group are the same.	2. Avoid qualifiers that reinforce stereotypes.	3. Identify people by identity characteristics only when relevant.	4. Be aware of language that, to some people, has questionable racial or ethnic connotations.
5. Be aware of the possible negative implications of color-symbolic words.	6. Avoid patronizing language and tokenism toward any racial or ethnic group.	7. Substitute substantive information for ethnic clichés.	8. Review media to see if all groups are fairly represented.

relevance these concepts have to public education by reviewing broad patterns in school performance data revealing an "opportunity gap" in mathematics education.

Patterns in data reveal mathematics education's opportunity gaps

The achievement gap is a starting place to understanding why diversity matters in the mathematics classroom. This concept describes the different levels of performance outcomes on mathematics standardized tests when comparing students with different racial/ethnic and social class backgrounds. For example, passing rates on mathematics test scores for white students are greater than that of Black/African American students. You may have heard of these differential scores and the phrase "achievement gap" in the news before; it is how journalists and politicians describe the situation. We will look at some examples of this from the Brookings Institute (a popular US think tank) before moving beyond the term "achievement gap" and thinking more accurately about the situation as an "education debt" and an "opportunity gap," two phrases put forth by well-known researchers in public education.

In the late 1990s, the achievement gap concept became a popular idea that forced politicians and society to confront the failure of public education for its students of color and students from lower socioeconomic class backgrounds. In 2001, the No Child Left Behind (NCLB) Act (as a reauthorization of the Elementary and Secondary Education Act) squarely addressed the equity issues in public education by requiring all states in the United States to have rigorous academic standards and for schools to be held accountable to improving student test scores for each subgroup of students, including when broken down by race, socioeconomic status, and students in special education. These addressed both literacy and mathematics standardized test scores. The exact mechanisms of holding schools accountable changed when the reauthorization took place again in 2015, ending NCLB and replacing it with Every Student Succeeds Act (ESSA) (2015). Throughout both policies, the goal was eliminating the so-called achievement gap for all schools in the United States.

As a well-known non-partisan think tank, the Brookings Institution provides insight into the achievement gap over time. In a more recent article (Hansen et al., 2018), they draw from standardized testing data over the years of NCLB to determine how much, if any, of the achievement gap was reduced. They use mathematics and literacy data from the National Assessment of Education Progress (NAEP) test, which is a sample of fourth- and eighth-grade students across the United States. Overall, they noted a steady decline in the racial/ethnic achievement gap but with a large differential still between white students and students of color. Although the gap has lessened, it still looms large. Unfortunately, during the NCLB years the gap between students from higher and lower social class backgrounds remained static. In answering their question about making any progress since NCLB, they write, "Though the early NCLB era did see some improvements in student achievement, it appears that this progress has largely stalled during the latter half of the period" (Hansen et al., 2018, n.p.). For example, in the year 2017, there still existed a difference of 0.74 standard deviation units between the raw scores of Black/African American and white students on the fourth-grade mathematics and literacy tests. The Brookings Institution interprets this to mean that, roughly speaking, white students are close to two years ahead academically.

These data at the national level are replicable in your local area. Nowadays it is easy to find performance data for schools across the United States, and they even now report data to indicate an equity rating for each school based on the differential outcomes between white students and students of color and other demographic breakdowns. Greatschools.org is a website that houses publicly available data on student performance (using state standardized test scores) for most schools in the United States. For each school, they have an "equity overview" review that compares the passing rate of standardized tests for white students to students of color, for students from lower socioeconomic status and higher socioeconomic status, and for students in special education programs to those that are not. They look at the data in total and provide summary statements as follows (based on one school as an example): "Underserved students at this school are performing better than other students in the state, though this school may still have achievement gaps." If you have not yet looked at data for schools in your area, this can prove worthwhile to understand your local context. Look at a few different schools in your region and especially make sure to note the performance data from a variety of different schools. You will find very stark differences in the data, with higher performance data associated with affluent schools with predominantly white student populations.

Educational researchers who are experts in thinking through this data make a very important point by moving us beyond the term "achievement gap" as we think about equity in public education. They ask us to consider the following questions: "Does a standardized test score truly measure a student's *achievement* as the result of school?" and "What are the causes and impacts of these differential outcomes?" Gloria Ladson-Billings is a celebrated educational researcher who developed the concept of culturally relevant pedagogy, a topic we will explore deeply in Chapter 4's focus on race. When she was president of the American Educational Research Association, Ladson-Billings gave an address to the organization in which she pushed directly against the term "achievement gap." First, she articulates the phenomenon as seen in the data with some extra attention beyond test scores:

> Even when we compare African Americans and Latina/os with incomes comparable to those of Whites, there is still an achievement gap as measured

by standardized testing… It also exists when we compare dropout rates and relative numbers of students who take advanced placement examinations; enroll in honors, advanced placement, and 'gifted' classes; and are admitted to colleges and graduate and professional programs.

(2006, p. 4)

With a more nuanced perspective of the issue at hand, Ladson–Billings suggests that we focus less on the year-to-year gains or losses in differential achievement and more on the ways that the gaps we see each year represent the accumulation of an "education debt" over decades that impacted lower social class students and students of color. She argues "that the historical, economic, sociopolitical, and moral decisions and policies that characterize our society have created an education debt" (p. 5). For the historical debt, she reviews African American's experience in the US public education system beginning with exclusion from education during enslavement that continued with poorly supplied schools during segregation. She also reminds us of the American Indian boarding schools with a stated mission to force assimilation that "belonged nowhere. The assimilated Indian could not fit comfortably into reservation life or the stratified mainstream" (p. 5). Similarly, "Latina/o students also experienced huge disparities in their education" (p. 6) documented as early as 1848. For the economic debt, Ladson–Billing reminds us of the funding inequities in public education. This took place in a straightforward manner when schools were segregated but still exists today with the various funding formulas and zoning of school districts that align to real estate markets. Finally, Ladson–Billings highlights that the education debt also includes the years that people of color in the United States have been excluded from political life. Because people of color in the United States were denied access to voting rights and had to fight for these during the civil rights era, "families of color have regularly been excluded from the decision-making mechanisms that should ensure their children receive quality education" (p. 7). Taken together, the histories of bad schooling, the economies of schools for people of color, and the exclusion of people of color in the politics of schools create a significant challenge for families of students of color in the present day. Heavily burdened by these accumulations of the education debt, it is a far cry to state that students of color and low-income students are failing to "achieve" like their white counterparts in the schools.

In this way, Ladson–Billings focuses us to look at the centuries-long contexts that create the so-called achievement gap that we see in public education. Similarly, another celebrated scholar of equity in public education, Richard Milner, pushes us to think about this context in the present day as an "opportunity gap." By shifting to the contexts required for effective learning to take place, Milner is not locating the problem in the students' lack of achievement but rather the societal and school-based scenarios that make the learning opportunities different for students. He writes:

> The opportunity Gap Framework covers five interconnected areas that I
> believe are essential to helping educators shed light on and address oppor-
> tunity gaps ... 1) rejection of color blindness; 2) ability, willingness, and
> skill to understand, build on, and work through cultural conflicts; 3) abil-
> ity and willingness to understand how the meritocracy myth operates; 4)
> ability and willingness to recognize, disrupt, and shift low expectations
> and deficit mind-sets; and 5) willingness to counter and rethink context-
> neutral mind-sets and practices.
>
> *(2020, p. 23)*

Milner's list suggests that these particular areas of classroom teaching and school
culture and practice are the reasons why we see differential performance outcomes for students. The opportunity to learn is stifled given major issues that
need to be interrupted and corrected by educators. We will look again at Milner's
concept in Chapter 4 when we consider his response to the opportunity gap,
what he calls "Opportunity-Centered Teaching."

Identity and diversity in mathematics education research

Alongside the political discourse of the so-called achievement gap and the more
careful thinking of equity in public education coming from Ladson-Billings,
Milner, and others, over the past 50 years scholars in mathematics education
have also dedicated efforts to thinking deeply about the impact of diversity in our
classroom spaces and mathematics teaching. Many times, these efforts focus on a
particular identity, such as race/ethnicity. Each of the remaining chapters in this
book presents examples of research findings specific to a facet of human diversity
in the mathematics classroom. As we work through these chapters, we'll keep
in mind what intersectionality tells us: facets of human diversity work together
at the individual and social level to create unique experiences as identities layer
together.

Scholars of mathematics education have announced these efforts in diversity,
one celebrated researcher Rochelle Gutierrez named them the "sociopolitical
turn" in the field. Grounded by opportunity gap data and a quest for better
equity in mathematics education, researchers have moved to understand more
deeply the social and cultural theories that inform student identities and thus
there classroom experiences. She writes:

> I use the term sociopolitical turn to reference the growing body of research-
> ers and practitioners who seek to foreground the political and to engage
> in the tensions that surround that work. The sociopolitical turn signals the
> shift in theoretical perspectives that see knowledge, power, and identity
> as interwoven and arising from social discourses. Adopting such a stance
> means uncovering the taken-for-granted rules and ways of operating that

privilege some individuals and exclude others. Those who have taken the sociopolitical turn seek not just to better *understand* mathematics education in all of its social forms but to *transform* mathematics education in ways that privilege more socially just practices.

(2013, p. 40)

Gutierrez notes that most researchers in this turn draw from the relevant concepts in sociology, anthropology, and political theory among a handful of other social science disciplines to consider best practices in equity for mathematics classrooms.

As we move into the subsequent chapters, each will have a focus on a particular facet of human diversity. We'll start each with a discussion of the facet of human diversity and draw from relevant social science concepts, similar to the work of mathematics education researchers. Next, we will look at major findings in general education scholarship as it relates to the specific facet of human diversity and third a review of major findings specific to mathematics classrooms. For example, the next chapter will focus on race and ethnicity. First, we will start broadly with a review of what the social sciences says about race. Next we will narrow the focus by connecting race and ethnicity to public education in the United States. Finally, we will focus specifically on race/ethnicity in mathematics classrooms. We'll follow the same approach for gender, social class, sexuality, ability, and language in the subsequent chapters. You will notice that several times throughout these approaches, an intersectional aspect to the work will come up as we isolate each identity for focused attention. With such isolation, we can learn important patterns in the data and research but we must also remain vigilant in approaching the work ahead with a commitment to the intersectional layering of human diversities in our classroom spaces.

Prompts and activities for discussion:

1. Because identities are socially constructed, we as individuals receive constant messages early on and throughout life that suggest how we should view specific identities. One activity that is suggested to help us to see this is the implicit association test. The point of this activity is not to place blame on individuals for their implicit associations, but for them to understand these subconscious habits of mind to better know our tendencies that might occur without such introspection. Take a few tests at the link below and discuss the questions that follow. Link: https://implicit.harvard.edu/implicit/takeatest.html.
    Questions:
    a) Did your implicit association surprise you?
    b) Reflect on the times when you were implicitly taught some of these habits of mind. How are these learned by images and ideas from the media, actions our family members take, experiences in school?
    c) Anticipate how this relates to mathematics teaching and learning. How would a teacher's subconscious habits of mind affect learning outcomes?

2. Learn more about what microaggressions look like by experiencing the implicit bias and microaggressions module developed by researchers at University of North Carolina: https://ready.web.unc.edu/section-1-foundations/module-4-implicit-bias-microaggressions/ Discuss the times you have noticed a microaggression take place, and, for the next week, have focused attention on microaggressions that take place in your world. After a week, come back together and share examples that you found now that you have increased awareness of this.

Classroom tips:

*   For each set of students you teach, develop a landscape of diversity narrative to describe their multiple identities. You can begin with the data that has already been reported by students' families into the demographic profile at your school system. Next, you might consider a "get to know you survey" in which you ask students for their preferred name and pronouns, and, as you deem appropriate for their age and readiness, you might be able to find a way to share any identities that they would like to with you. This will help you to determine the specific identities in your classroom; however, you can also assume some of the less visible identities in your classroom, like students from lower social class backgrounds or marginalized sexualities are present, even if you do not know that they are. Finally, never make any assumptions about one's identity based on your perceived impressions.
*   Pay close attention to your thoughts and actions as you work with students who have different identities than your own. Did a subconscious prejudice come to your mind as you were teaching a student mathematics or reacting to their behavior? Did you take a misstep by enacting some type of microaggression? The first step is to increase your own awareness of when these thoughts or actions come up for you; the next step is to eliminate actions that you take based on implicit bias or prejudice.

TABLE 3.2 Important terms and concepts in this chapter

Facets of human diversity	Each identity group or aspect within the diversity of the human population; a typical list includes the following facets: race/ethnicity, social class, gender, sexual orientation, ability, language, geography, religion, marital/familial status
Social construction	Each facet of human diversity has been created socially by people over time; there is no biological basis for each identity; however, an individual's identity is not a choice; even though they are socially constructed, identities are highly significant to a person's social experience in the world

(*Continued*)

TABLE 3.2 (Continued)

Marginalized or minoritized	The underrepresented identity for a given facet of human diversity, for example Black or African American, is a marginalized identity within race/ethnicity, whereas white is the dominant group (not marginalized)
Bias-free communication	Language practices that aim to eliminate biased ways of speaking and writing about marginalized identities; constantly evolving language practices; a best practice is to consult guidelines on a consistent basis to update your language practices
Implicit bias	Preconceived notions, such as stereotypes, about a particular identity within a facet of human diversity often the result of influences by society, often subconscious
Discrimination	When an individual acts on their implicit bias or prejudice to cause harm to an individual from a marginalized identity group
Microaggression	Subtle acts of discrimination through interpersonal interactions; someone from a dominant identity group makes a simple comment or statement to someone from a marginalized identity group that, for example, reinforces a stereotype or dismisses an experience of discrimination; experienced often by those in marginalized identity groups with accumulating effects of personal trauma
Systemic oppression	Oppression of a marginalized or minoritized group at the institutional level, such as repeated patterns of racial discrimination in housing, the penal system, and public schools
Achievement gap	Phrase typically used in media and by politicians to describe the different performance levels by white students and students of color as well students from higher and lower social class backgrounds
Opportunity gap	A more fitting phrase to describe the achievement gap; points to the differences in equitable learning opportunities for students with the public education system
Education debt	The years of historical, political, and economic marginalization experienced by people of color that results in the inequitable public education system that we have today
Sociopolitical turn	In mathematics education, greater attention is paid to facets of human diversity to understand how to create equitable opportunities to learn in our classrooms; draws from social science concepts on diversity and identity to best address the issue

Further reading

Anti-Defamation League. (2014). Guidelines for achieving bias-free communication. https://www.adl.org/resources/tools-and-strategies/guidelines-achieving-bias-free-communication

Barroso, A., & Brown, A. (2021). Gender pay gap in U.S. held steady in 2020. *Pew Research Center.* https://www.pewresearch.org/fact-tank/2021/05/25/gender-pay-gap-facts/

Baumgartner, F., Epp, D., & Shoub, K. (2018). *Suspect citizens: What 20 million traffic stops tell us about policing and race.* Cambridge University Press.

Gutierrez, R. (2013). The sociopolitical turn in mathematics education. *Journal for Research in Mathematics Education 44*(1): 37–68.

Hansen, M., Levesque, E.M., Quintero, D., & Valant, J. (2018). Have we made progress on achievement gaps? Looking at evidence fom the new NAEP results. Brookings Institution. https://www.brookings.edu/blog/brown-center-chalkboard/2018/04/17/have-we-made-progress-on-achievement-gaps-looking-at-evidence-from-the-new-naep-results/

Ladson-Billings, G. (2006). From the achievement gap to the education debt: Understanding achievement in U.S. schools. *Education Researcher 35*(7): 3–12.

May, V.M. (2015). *Pursuing intersectionality, unsettling dominant imaginaries.* Routledge.

Milner, R. (2020). *Start where you are but don't stay there, 2nd ed: Understanding diversity, opportunity gaps, and teaching in today's classrooms.* Harvard Education Press.

Riegle-Crumb, C., & Humphries, M. (2012). Exploring bias in math teachers' perspections of students' ability by gender and race/ethnicity. *Gender and Society 26*(2): 290–322.

4

A WHITE INSTITUTIONAL SPACE

Race and mathematics education

White supremacy in the United States has a deep and sustaining history, early on with the enslavement of African Americans and genocide and forced removal of indigenous people throughout the land and continuing nowadays with clear patterns in economic and sociopolitical data revealing discrimination for people of color in this country. Over centuries, we have made great progress in shifting toward greater civil rights and equity, drawing from the variety of groups and individuals in the civil rights movement from Dr. Martin Luther King Jr. to the Black Panthers and today's Black Lives Matter movement, but there is more work to be done. As a major feature in today's society, white supremacy appears in the mathematics classroom as well. Black mathematics educators like Danny Martin assert that mathematics education, and even specifically groups like the National Council of Teachers of Mathematics (NCTM), are a racialized project conceived of by whites and in their interest, and one that only superficially takes efforts toward addressing racial injustice as experienced through mathematics teaching and learning. In this chapter, we will confront how mathematics education perpetuates white supremacy and in what ways we can teach critically to interrupt this. As with the remaining chapters in this book, we'll first start with describing the broader nature of white supremacy as a social construction, move to what is known about racism in the US public education system, and then narrow our focus to mathematics teaching and learning.

White supremacy and the social construction of race

As we start our work in understanding white supremacy and racism, we can think about using bias-free communication. Throughout this text, I will use the phrases people of color, students of color, and Black, Indigenous, people of

DOI: 10.4324/9781003322566-4

color (BIPOC) as inclusive phrases to describe people who are not of European ancestral origins. BIPOC is a newer and increasingly used term throughout education research and discussion. Within this broad category, we have several identity markers and most everyone prefers a more specific term to describe themselves. Most importantly, when working across divergent identities do not assume that you know how the person identifies. For example, many in the United States prefer the term "Black," whereas some might prefer "African American." Hispanic is an ethnic identity referring specifically to Spanish speaking cultures, whereas a Latino/a/x individual is asserting their identity as someone with Latin American cultural heritage. Several within this community use Latinx to describe the whole social group (rather than Latinos, for example) for its inclusion of all gender types. Indigenous people in the United States sometimes prefer an identity with their particular tribe, or the term "American Indian," or Native American, as some examples. AAPI is an acronym for Asian American and Pacific Islander people who live in the United States.

Another important first step as we begin our work is to assert our racial and ethnic identities to position us within the present-day racialized experience. You may want to pause and think about your own racial identity and what it means for you and the work we'll do together before reading my own positionality. Here is mine: I am a white person who recognizes that white supremacy presently exists. This means first that I acknowledge the privilege that I receive by living in a white body and, when I can, use that privilege to teach other's about white supremacy and otherwise interrupt it. Through years of study and experience, I have faced race and racism more than most white people do. Doing so, I know I can never have a complete understanding of the BIPOC experience but I can learn about it as much as I can. I never push a BIPOC friend or acquaintance to describe their experiences to me; they are not responsible for my own education in white supremacy. I "do my homework" by reading authors and reading news, looking up statistics about race and racism. As a white person, I support movements like Black Lives Matter but do not lead in them.

As with all identities, race is a social construction. A casual experiencer of the modern world will no doubt have some basic understanding of or experiences with the notion of race. However, after this, experiences with race and racism differentiate rather starkly. Broadly speaking, some people (typically BIPOC) consciously experience racism every day and others are subconsciously a part of racism but can decide to consciously engage with the concept of race or not. Take the latter, and I am referring to the white population. Whites have racial experiences ranging from the following: those who choose to have a critical, keen awareness of race and racism, to those with an attitude/experience of colorblindness, to those acting along the lines of (and sometimes admitting to) white supremacist ideology. In what follows, we consider the contributions of three scholars, the first two describing how the construction of white supremacy came to be and the last contributing in words the social experience of being a person of color in a racist society. Although the term "white supremacy" might conjure up the image of neo-Nazis protesting against, say, the Black Lives Matter movement, we are using it here to specifically denote how race as a social construct functions as a sociopolitical ordering of relations between groups of people. The phrase accurately describes the hierarchy contained with this ordering, with whites on top and other races and ethnicities marginalized and minoritized.

Historian George Fredrickson, author of *White supremacy: A comparative study in American and South African history* (1981), studies how this habit of mind historically *came to be*, making the claim that the developmental history of white supremacy in the United States was entangled with economic opportunity and sociopolitical development. On the one hand, white supremacy came to be because European settlers in the United States south needed justification, a reason, to enslave people of African descent, and continued long after the emancipation of slaves in 1865. Fredrickson is careful to note, however, that economics did not entirely drive racial divides. Instead, the rise of ethnic consciousness and economic advantage fed off each other in the development of racial and class distinctions over time.

Another scholar, scientist Stephen Jay Gould, provides a complement to Fredrickson's study on the historical development of white supremacy. *The Mismeasure of Man* (1996) tells the chronology of the mathematicians and scientists who attempted to validate white supremacy for its sociopolitical and economic needs. Essentially, it was a quest in biological determinism or to prove that "shared behavioral norms, and the social and economic differences between human groups—primarily races, classes, and sexes—arise from inherited, inborn distinctions and that society" (p. 52). Over the years, efforts to document differences in brain size and certain styles of psychological testing all claimed to scientifically document white supremacy. One by one, Gould picks apart these studies for their flaws, from nineteenth-century studies of brain size to the 1990's *Bell Curve* (Herrnstein & Murray, 1990) argument relying on IQ testing.

To understand race as a social construct is to take a hard look at its development, as these two have done. By saying that "Race is a social construct," it means that the racial categories have no scientific viability. As documented in the histories above, the superiority of the white race was the result of historical developments in geographies, politics, and economics that necessitated claims of superiority with respect to ancestry and skin color. Saying that race is a social construct is not the same thing as saying "Race does not exist." Because it is a social construct, it has sustained effects on people and how they interact. We are all affected by race and racism, and it is difficult to escape, especially by simply stating "I am not a racist." It requires checking the habits of mind and actions resulting from societal influences that perpetuate white supremacy. For example, take a mathematics teacher who teaches mathematical reasoning to his white students and mathematical mechanics to his Black students. This teacher might be acting on an unchecked, subconscious attitude of white supremacy causing him to expect higher-order thinking from his white students and lower-order thinking from his Black students.

Any discussion of race and white supremacy should focus on the BIPOC experience. A classic text by W. E. B. DuBois, written in 1903, is *The Souls of Black Folk*, which describes Black consciousness in the United States. He opens the first chapter of the book by describing what it feels like "to be a problem" and describes the Black psyche as follows:

> After the Egyptian and Indian, the Greek and Roman, the Teuton and Mongolian, the Negro is a sort of seventh son, born with a veil, and gifted with second-sight in this American world,—a world which yields him no true self-consciousness, but only lets him see himself through the revelation of the other world. It is a peculiar sensation, this double-consciousness, this sense of always looking at one's self through the eyes of others, of measuring one's soul by the tape of a world that looks on in amused contempt and pity. One ever feels his twoness,—an American, a Negro; two souls, two thoughts, two unreconciled strivings; two warring ideals in one dark body, whose dogged strength alone keeps it from being torn asunder.
>
> *(p. 11)*

DuBois' description of twoness points to the strong identity that race places on BIPOC. A white teacher who states, "When I see people, I do not see color," simply showcases that white folks cannot know and feel what it is like to have a racial identity. Quite the contrary, a BIPOC person in a racist society is always conscious of their racial identity. This statement that a teacher is making is called "colorblind" and has been actively rejected by teachers and education researchers who work toward racial equity in schools. If you refuse to see a student's racial identity, you refuse to engage in the totality of their social experience which unfortunately includes significant discrimination.

Race and education: Critical race theory and culturally relevant and sustaining pedagogies

This section introduces you to some practices and theories in education that take up the notion of white supremacy and racial constructs directly. We begin with an understanding of critical race theory and move to examples of practice from culturally responsive and sustaining pedagogies. We will also review Milner's "opportunity-centered teaching" as his response to the opportunity gap we discussed in Chapter 3.

You may have heard about Critical Race Theory in the news lately. Unfortunately, partisan politics in the Unites States have caused the concept to become a bit confused and entangled with other ideas. It is not a theory that we teach our students in K–12 public schools, but it is a theory that helps us to understand how racism plays out in the public school system. Critical race theory emerged as a field within legal studies and has since been taken up by a great many critical education scholars and practitioners. Gloria Ladson-Billings and William Tate provide an introduction to its application to schools with the article "Towards a critical race theory of education" (1995). Keeping in the critical race theory tradition, the following three tenets are put forth: "1) Race continues to be a significant factor in determining inequity in the United States. 2) US society is based on property rights. 3) The intersection of race and property creates an analytic tool through which we can understand social (and, consequently, school) inequity (p. 48)." Many educational theorists and practitioners have applied critical race theory to critique mainstream education and to imagine anti-racist schooling experiences.

In other words, inequities in property holdings coexist with inequities in power. Connecting this to education, Ladson-Billings and Tate suggest property's relevance to unequal schooling opportunities. For example, where a student goes to school is determined by her family's residence and this affects the type of schools she attends. Although official desegregation occurred in the 1960s, de facto segregation occurs via "white-flight" as a particular school district's community changes racial demographics. Other processes and structures related to property and real estate have also led to segregated schools, such as redlining, the practice where particular neighborhoods exclude people of color by preventing mortgages for the purchase of homes. Such analysis runs counter to common assumptions about schooling; public education is assumed to level the playing field by providing all with equal opportunity to life's chances.

It is not enough to suggest that the suburban-urban divide, tracking within diverse schools, and the like, occurred as the result of economic divisions within society. Ladson-Billings and Tate write:

> If racism were merely isolated, unrelated, individual acts, we would expect to see at least a few examples of educational excellence and equity together in the nation's public schools. Instead, those places where African Americans do experience educational success tend to be outside the public schools. Some might argue that poor children, regardless of race, do worse in school, and that the high rate of poverty among African Americans contributes to their dismal school performance; however, we argue that the cause of their poverty in conjunction with the condition of their schools and schooling is institutional and structural racism.
>
> *(p. 55)*

Ladson-Billings and Tate carefully argue against any notion that poverty in its entirety causes the poor educational outcomes for BIPOC.

Moving from analysis and toward general pedagogies that interrupt racial injustice, those within the critical race education camp take on a variety of stances. The first of these we consider, culturally relevant pedagogy, comes again from Gloria Ladson-Billings. She provides a three-faceted approach to teaching BIPOC students that aims to increase student success while maintaining dignity and respect for one's culture. The three facets are (1) academic success, (2) cultural competence, and (3) critical consciousness, and I include her introductions of each from her initial article on culturally relevant pedagogy "But that's just good teaching: The case for culturally relevant pedagogy" (1995). Academic success:

> Despite the current social inequities and hostile classroom environments, students must develop their academic skills. The way those skills are developed may vary, but all students need literacy, numeracy, technological, social, and political skills in order to be active participants in a democracy.
> *(p. 160)*

All educational measures provided to BIPOC students should adequately prepare them for economic circumstances as they are today. Second, cultural competence: "Culturally relevant teaching requires that students maintain some cultural integrity as well as academic excellence. In their widely cited article, Fordham and Ogbu (1986) point to a phenomenon called "acting white," where African American students fear being ostracized by their peers for demonstrating interest in and succeeding in academic and other school related tasks" (pp. 160–161). Culturally relevant pedagogy provides education for students of color where conscious attempts are made to embrace and not reject one's own culture. Finally, critical consciousness:

> Culturally relevant teaching does not imply that it is enough for students to choose academic excellence and remain culturally grounded if those skills and abilities represent only an individual achievement. Beyond those individual characteristics of academic achievement and cultural competence, students must develop a broader sociopolitical consciousness that allows them to critique the cultural norms, values, mores, and institutions that produce and maintain social inequities. If school is about preparing students for active citizenship, what better citizenship tool than the ability to critically analyze the society?
> *(p. 162)*

This facet of culturally relevant teaching directs teachers to expose children to injustice in the world.

Django Paris provides us with culturally sustaining pedagogy as a stronger assertion of Ladson-Billings' second and third tenets of culturally relevant pedagogy. Paris is concerned that too many interpret culturally relevant pedagogy as simply accommodating student culture into the dominant curriculum without truly engaging BIPOC culture in the classroom. He writes:

> We must ask ourselves if the research and practice being produced under the umbrella of cultural relevance and responsiveness is, indeed, ensuring maintenance of the languages and cultures of African American, Latina/o, Indigenous American, Asian American, Pacific Islander American, and other longstanding and newcomer communities in our classrooms....I offer the term *culturally sustaining pedagogy* as an alternative that I believe embodies some of the best past and present research and practice in the resource pedagogy tradition and as a term that supports the value of our multiethnic and multilingual present and future. The term *culturally sustaining* requires that our pedagogues be more than responsive or relevant to the cultural experiences and practices of young people—it requires that they support young people in sustaining the cultural and linguistic competence of their communities while simultaneously offering access to dominant cultural competence.
>
> *(2012, pp. 94–95)*

In Chapter 3, we considered Richard Milner's description of the opportunity gap in public education today. As a response, he provides four tenets to his "opportunity-centered teaching" as practices and habits that teachers can take to especially address the lack of opportunities in the system perpetuating inequities for BIPOC students in public education. These tenets are provided in Table 4.1 on page 57 with a salient quotation from his description of each.

As we move into our discussion on mathematics education and race, consider the possibilities that these three scholars have presented. First, how is mathematics education set up as a racialized space? Second, how do we teach to interrupt this space? What are examples of culturally relevant and sustaining pedagogies that we can employ in the mathematics classroom? Finally, what are specific considerations for varying racial and ethnic identities that appear in our diverse classrooms?

What is white mathematics education and how to interrupt it

So far in this chapter we have looked broadly at race and white supremacy in society and next applied this to education in general. Here, we take the next step in thinking about race and our efforts in teaching mathematics with a critical race perspective. We start with a look at the racialized nature of mathematics education policy and distinctions among those who discuss race in mathematics education. Next we look at several considerations and projects aiming to increase access and equity with respect to mathematics education. These contributions correspond directly to the theory and approaches laid out in the previous sections, including critical race theory and culturally relevant teaching.

Danny Martin, a professor of mathematics education, provides clear discussions about the racialization of mathematics education policy in the United States. His article "Race, racial projects and mathematics education" (2013) foregrounds the construct of race in understanding the historical development of math teaching and learning, at least as it's talked about at the top level. He argues how mathematics education can be characterized as "white institutionalized space." This means that white actors have dominated the definition and purpose of mathematics education over the course of its development. They have and continue to hold math education's positions of power. Finally, mathematics education has long been portrayed as neutral, impartial, and unconnected to race and white supremacy. The last is perhaps the easiest to see firsthand. There has been a long tradition to actively state that mathematics is not relevant to race, and yet race is so clearly related to student performance in it. And yet, how often do we say and hear claims of math as the "universal language," the objective, value-free knowledge? In this sense, we might say that mathematics education, as it has been conceived, should be more accurately called white mathematics education.

A compelling example of mathematics education as white institutional space is the tracked experiences that students endure in their schooling experiences. Students are placed according to their performance on standardized tests and teacher recommendations. The research article by Valerie Faulkner and her colleagues at North Carolina State titled "Race and teacher evaluations as predictors of algebra placement" (2014) sheds insight into how the tracked experiences entangle with race and implicit biases. Essentially, their study reports "Black students had reduced odds of being placed in algebra by the time they entered 8th grade even after controlling for performance in mathematics" (288). The study calls our attention toward the impact of teacher recommendations for placement, and the researchers are careful in suggesting that such outcomes point to implicit bias in a white supremacist society, rather than conscious actions taken by individual teachers (although this certainly does happen).

Moving to interruption of these problematic patterns via best practices in equitable mathematics instruction, Rochelle Gutierrez focuses on successful mathematical education for students of color. "Advancing African-American, urban youth in mathematics" provides a case study of one mathematics department who consistently advanced urban students of color into higher-level mathematics courses. In her study, Gutierrez attributes this to "five characteristics of the department: a rigorous curriculum and the support to maneuver through it; active commitment to students; commitment to a collective enterprise; a resourceful and empowering chairperson, and standards-based instructional practices" (p. 63). She found the department to embrace the tenets of culturally relevant pedagogy as well, as discussed in the earlier section.

Complementing this, the article "Communities for and with Black male students," by C. C. Jett, David Stinson, and Brian Williams presents four strategies found to be consistent among successful teachers of African American males. As they suggest, effective mathematics teachers for students of color are those who

> develop caring relationships that reach beyond the classroom, access and build on out-of-school experiences and community funds of knowledge during instruction, implement culturally relevant pedagogy throughout instruction, and disrupt school mathematics in particular and mathematics in general as a white institutional space.
>
> *(p. 286)*

Consistent application of Ladson-Billings' culturally relevant pedagogy and Milner's opportunity-centered teaching are seen here, and by this we are encouraged to think deeply about what this looks like in the mathematics classroom.

A review article on the variety of research efforts made with respect to interrupting mathematics education as a white institutional space comes from the article "Rethinking teaching and learning mathematics for social justice from a critical race perspective" (2016) by Gregory Larnell, Erika Bullock, and C. C. Jett. In the article, the authors position two definitive origins for the work of interrupting white mathematics education: Marilyn Frankenstein and Eric (Rico) Gutstein's work that applies Paulo Freire's pedagogy to mathematics and Bob Moses' "Algebra Project" aiming to increase the success in algebra by students of color (p. 20). Larnell et al. do not present these as two opposing perspectives that we must debate on and decide which will lead to the greatest interruption of white mathematics education. Rather, these are two approaches that resonate with the broader conversations regarding race and education, such as the contributions of Lisa Delpit and bell hooks in discussions about equity in literacy education. On the one hand, we want to teach students what they need to sustain cultures and deepen understanding of societal inequities. On the other, we need to maintain the commitment to provide our students with strong education in the dominant curriculum so they have access to opportunity through their colleges and career.

Let's say that a high school math teacher in an urban school teaches juniors and seniors of color and aims to interrupt white mathematics education. He entirely replaces a white mathematics education program (and its learning objectives) with a course of study rich in the inquiry of injustice, perhaps through statistical investigations of local circumstances or the mathematical modeling of social relations. This no doubt interrupts white mathematics education and demonstrates to students the power of mathematical thinking. However, the teacher might not be preparing his students for college-level mathematics requirements that are stuck in status quo proficiencies in algebra. This circumstance, where many students are prevented a college degree because of their lack of proficiency in mathematics, is a problem that we can work to change but for the time being is firmly in place. Thus, both providing greater access to white mathematics education, what we might call mathematical empowerment, and opening minds to mathematics beyond white mathematics education, what we might call revolutionary mathematics, are equally important aspects to the interruption of white mathematics education.

Additional mathematics education research focuses our attention on other racial identities including Latinx, indigenous, and AAPI students. As an example of strong contributions related to the Latinx student population, Civil (2009) reflects on her work with parents and communities as she develops a funds of knowledge approach to mathematics teaching akin to Milner's tenet #3. She describes this as work taking place in

> Mexican/Mexican American, working class communities in the Southwestern United States. Our efforts have been geared towards the development of learning environments that build on the students' and their families' knowledge and skills. But, how do we uncover that knowledge and those skills that families have? Through the Funds of Knowledge for Teaching project... teachers visited the homes of some of their students.
>
> *(p. 10)*

Recent research on indigenous students in mathematics classrooms also aims to engage with community knowledge and culturally sustaining pedagogy. Garcia-Olp, Nelson, and Saiz (2022) provide their curricular project title "IndigiLogix: Mathematics, Culture, Environment", which is taught to pre-college indigenous students in the United States. They decolonize mathematics learning by grounding curriculum in indigenous ways of knowing: "belonging, mastery, independence, and generosity" (p. 1). Although both approaches to Latinx and indigenous mathematics education involves culturally sustaining pedagogies and curricular convergence, it is important to note that community knowledges will look different for the student population that a teacher has. Engaging with the community will be an ongoing project for any teacher working across different racial and ethnic student identities. Reading these research studies can indicate methods for mathematics teachers to understand their local and specific cultural knowledges that will be an asset for teaching in the classroom.

For Asian American and Pacific Islander mathematics students, their challenges in the classroom are understudied and yet significantly complex. One aspect of great urgency is the racism that exists within our field, essentializing all AAPI students as a "model minority"; individual AAPI students are assumed to outperform in our classrooms. Wu and Battey (2021) provide recent research to unpack how participants in their study, as Chinese and Taiwanese American students of mathematics, confronted the "model minority" narrative in our discipline. We must remember that AAPI is a panethnic group with a variety of different populations, and of course every individual within a group will uniquely perform in a mathematics classroom as per a host of other factors. Wu and Battey remind us that

> Framings of Asian Americans as better prepared mathematically are broad generalizations in and of themselves. For example, in averaging across Asian Americans to view them as racially outperforming other groups, the framings ignore subgroups such as Hmong and Filipino students who typically are not served well mathematically.
>
> *(p. 582)*

Notice that their clear description of these nuances reflects the social science perspective that patterns of outperformance in any particular group is related to how well the population is served by its educational opportunities, not by inherent or biological differences.

Most generally, our work as mathematics teachers needs to prioritize affirming racial and ethnic identities in our classrooms and engaging in full understanding of how this comes to bear on our teaching efforts. On the one hand, we can gain greater information about the cultural assets that these identities bring to our classrooms and thereby deepen cultural competence and BIPOC representation in mathematics, using a funds of knowledge approach. We also engage in the ways that particular racial and ethnic identities foster stereotyped expectations for learning mathematics, such as the "model minority" for AAPI students and the lowered expectations for learning higher-order mathematics that BIPOC students experience.

Activity for discussion:

1. The following is a semi-fictional case study of mathematics teaching. Read and discuss how this case does or does not interrupt white mathematics education. Does the case correspond to any of the following: culturally relevant teaching, culturally sustaining education, or opportunity-centered teaching?

Mr. Jones is a white algebra teacher working in an urban/suburban school in the northeast United States. The student population is approximately 60% African American and 35% white. 40–50% of the students qualify for free or reduced price lunch.

In this district, students are tracked in classes based on test scores and teacher recommendation. This case focuses on Mr. Jones' lowest level Algebra 1 class. There are 18 students total: 15 African American, 3 white, 13 boys, and 5 girls. In the last two weeks of May, the district curriculum requires this class to review number plots and learn about measures of central tendency (mean, median, mode) and spread (range).

After students finish their test on the previous unit, Mr. Jones begins this unit with the remaining ten minutes of the class period. He holds a whole-class conversation about the students' preferences in popular music; it becomes clear that most students like hip-hop. Although some students may prefer R & B to hip-hop, all students in the class have great familiarity with the hip-hop genre. He then assigns students to bring in a favorite hip-hop song to share with the class the next day and requires that the songs are free of explicit lyrics.

In the next lesson, Mr. Jones begins with a think-pair-share: "What makes some hip hop songs better than others?" At the opening activity's conclusion, Mr. Jones summarizes that the class thinks songs can be rated on the following: beat, flow (of lyrics), catchy hook, lyrics theme, and overall rating.

The students next use these categories to rate six of the songs that were brought to class that day. Students record their ratings on their sheets and put them on the board. Mr. Jones prompts students to think how they should display the data. After a brief whole-class discussion, the class decides that a number plot (which they learned earlier in the year) would work best. The day's lesson concludes with students recording data on number plots.

The next day, Mr. Jones asks students to discuss the data in groups. He prompts them for productive discussion: "What song did we seem to like the best? How can you tell that?" He also prompts students to think this through for each of the different categories (beat, flow, etc.). Students excitedly discuss and debate the answer to these prompts. Some have relevant prior knowledge and suggest finding the "average rating" for each song. Other students ask about how to do this and are taught by each other, with Mr. Jones facilitating.

In the ensuing whole-class discussion, a debate emerges. Roughly half the class thinks Song A is the class favorite because its average is highest. But some students, who happen to like Song C more, are arguing that Song C is the class favorite because the most common score it received was a 5/5 for the overall rating. Throughout this and previous conversations, Mr. Jones and the class also debate the themes of the six songs. Some of the students in the class openly disagree with a few songs' themes: emphases on "gang-life," consumerism, and/ or treatment of women. Mr. Jones facilitates discussions so that students' musical preferences are simultaneously honored and problematized.

At this point, Mr. Jones interjects with some specific points to make. He congratulates the class for identifying the need to compute averages and for helping each other remember how to do this. He also highlights their discovery of another important feature in the data: looking for the value with the **most** responses. He directly teaches that these two **statistics** are referred to as **mean** and **mode**.

In a similar fashion, Mr. Jones prompts students to discover the concepts of **median** as well as **range**, a measure of spread. After a week and a half, Mr. Jones is satisfied that the students have learned the required concepts of the unit as well as practiced them on the hip-hop and other data sets. Because these objectives are finished within less time than allotted for the unit, Mr. Jones pushes the students to think more about measures of spread in the following final two days of instruction.

Mr. Jones says to the class: "You have noticed that for some songs we agreed quite a bit but for others we disagreed. Which song did we agree the most on? How do you know?" In small groups, students use the range but are prompted by Mr. Jones to think about computing a "typical distance from the mean." Ultimately, students discover a method to average these distances. In a whole-class discussion, students share their methods and Mr. Jones clarifies how to compute **absolute deviation**.

In the final instructional day, Mr. Jones reviews absolute deviation, both how to compute it and what it means. He shares with the students that this is material deemed too advanced for this class, and asks if they want more. Given their sustained enthusiasm, he next shares with students how to compute **standard deviation** for their hip hop data. As they complete this, he reminds the students that standard deviation tells us the "average distance from the mean" in a similar way to absolute deviation.

At the conclusion of the unit, Mr. Jones is satisfied with his students' mastery of district curriculum as well as advanced material. In final discussions of the unit, students continue to reflect on their more nuanced understanding of hip hop music. Students continue to debate themes in hip-hop as well as what makes a song sound good.

Classroom tips:

- Identify your own racial and ethnic identity and recognize how this informs your experience in the racialized world. Consider how you approach students from other racial and ethnic identities.
- Avoid colorblind thinking and recognize your students' racial and ethnic identities as you work with them. Notice if subconscious racial biases come to your mind and prevent yourself from acting on these biases, for example eliminate any tendency to exert greater discipline over BIPOC students.
- Learn as much as you can about your students' cultures and their communities. Do not burden your students by inviting yourself over to their homes but immerse yourself in the community in non-intrusive ways, such as attending community events and outdoor public gatherings and cultural sites. Make yourself known to your students' community; approach these opportunities with a mindset to want to learn more about the funds of knowledge that you can engage in your classroom.
- Develop unit plans with key questions and motivations that engage student culture in sustaining ways. Use the concepts from culturally relevant and sustaining pedagogies and opportunity-centered teaching to ensure that your units are academically rigorous but extend students' knowledge about their own cultures and worldviews.

TABLE 4.1 Milner's opportunity-centered teaching

Tenet #1: opportunity-centered teaching is about relationship cultivation	"Are teachers prepared and willing to build relationships with students whose life realities are very different from theirs? Are teachers prepared and willing to build relationships with students who look, act, and interpret the world differently from their teachers? Are teachers willing to relax their egos and view students – all students – as young people worth fighting for?" (p. 222)
Tenet #2: opportunity-centered teaching is about building community knowledge to inform practice	Many educators might think negatively about BIPOC communities; however, "every single student in our schools, including those who are often placed on the margins of learning, should be viewed as a vessel of knowledge and knowing: all of these young people come to schools with significant intellectual and cultural gifts and talents... Building community knowledge involves educators deliberately learning about their community and building relationships" (pp. 234–235)
Tenet #3: opportunity-centered teaching is about curriculum convergence	"What we sometimes forget is that adolescents' outside-of-school interactions and involvement can be viewed as part of a curriculum, albeit 'extra.' What if schools better constructed inside-of-school curriculum and learning opportunities with those that are outside of the regular curriculum? How could we build stronger synergy between the formal curriculum practices of school and students' practices and engagement outside of the regular school day?" (p. 240).
Tenet #4: opportunity-centered teaching is about psychological and mental health	"The mostly White teaching force can develop skills to recognize the ways in which racist acts can have real consequences for students' mental, emotional, and psychological health. Also, teachers can develop tools to help disrupts practices that play a role in incidents of racism experienced by students of color in the classroom. Thus, teachers should develop the skills to identify the ways in which they perpetuate inequity and racism and how they contribute – albeit unintentionally – to students' socioemotional and psychological strain" (p. 248).

TABLE 4.2 Important terms and concepts in this chapter

White supremacy	A historical, political, and economic circumstance resulting in whites as the dominant race with other races as marginalized
BIPOC	An umbrella term standing for Black, Indigenous, and People of Color
Critical race theory	Used by educators to explain the historical and sociopolitical circumstance that results in the inequitable schooling experiences for BIPOC students
Culturally relevant and sustaining pedagogies	Research in education exemplifying best practices for teaching BIPOC students; stresses the importance of teaching students with academic rigor and to deepen knowledge of student culture
White institutional space	Because mathematics education was conceived as a space by white elites and for white students, until we address race and culture in our classrooms it continues to perpetuate an educational program that works for the dominant racial and cultural white experience
Funds of knowledge, asset-based thinking	An approach to working with BIPOC students that prioritizes students' cultures as a knowledge to extend as opposed to something standing as an obstacle to learning the dominant curriculum (deficit thinking)

Further reading

Civil, M. (2009). A reflection on my work with Latino parents and mathematics. *Teaching for Excellence and Equity in Mathematics* 1(1): 9–13.

DuBois, W.E.B. (1994). *The souls of black folk.* Dover Thrift.

Faulkner, V.N., Stiff, L.V., Marshall, P.L., Nietfeld, J., & Crossland, C.L. (2014). Race and teacher evaluations as predictors of algebra placement. *Journal for Research in Mathematics Education* 45(3): 288–311.

Fredrickson, G. (1981). *White supremacy: A comparative study in American and South African history.* Oxford University Press.

Garcia-Olp, M., Nelson, C., & Saiz, L. (2022). Decolonizing mathematics curriculum and pedagogy: Indigenous knowledge has always been mathematics education. *Educational Studies* 58(1): 1–16.

Gould, S.J. (1996). *The mismeasure of man, Revised and expanded.* Norton.

Gutierrez, R. (2000). Advancing African-American, urban youth in mathematics: Unpacking the success of one math department. *American Journal of Education* 109(1): 63–111.

Jett, C., Stinson, D., & Williams, B. (2015). Communities for and with black male students. *The Mathematics Teacher* 109(4): 284–289.

Ladson-Billings, G. (1995). But that's just good teaching: The case for culturally relevant pedagogy. *Theory Into Practice* 34(3): 159–165.

Ladson-Billings, G., & Tate, W. (1995). Toward a critical race theory of education. *Teachers College Record* 97(1): 47–68.

Larnell, G., Bullock, E., & Jett, C. (2016). Rethinking teaching and learning mathematics for social justice from a critical race perspective. *Journal of Education 196*(1): 19–30.

Martin, D. (2013). Race, racial projects, and mathematics education. *Journal of Mathematics Education 44*(1): 316–333.

Paris, D. (2012). Culturally sustaining pedagogy: A needed change in stance, terminology, and practice. *Educational Researcher 41*(3): 93–97.

Wu, S.Y., & Battey, D. (2021). The cultural production of racial narratives about Asian Americans in mathematics. *Journal for Research in Mathematics Education 52*(5): 581–614.

5

SOCIAL CLASS HIERARCHIES AND MATHEMATICS EDUCATION

To reproduce or interrupt?

How does today's socioeconomic order affect education? Should mathematics education allow for social mobility or reproduce class structure? Are students with different social class backgrounds afforded the same or different mathematics education? We will explore these questions throughout this chapter to reveal that traditionally mathematics teaching does not promote social mobility as much as we think because most of the time public education "reproduces" the social classes. This means that students from a working-class background, for example, will likely be prepared for and end up in working-class jobs and income levels as adults. Mathematics teaching plays a role in this unfortunate reality, and as mathematics teachers we can interrupt the pattern with equitable practices in our classrooms. As with the previous chapter on race, we open with broad social theory, this time on social class, and move next to its application to education and finally to mathematics education.

Critically understanding capitalism and social class hierarchies

In this section, we look to social class as a social construct that emerges due to the logics inherent to capitalist economic order. This begins with general critiques of capitalism, from the classic text of Karl Marx to contemporary scholar David Harvey. Overall, the criticism causes us to reconsider what we are told to believe: capitalism is a fair competition with winners and losers. On the contrary, capitalism requires a social class hierarchy (executive, affluent professional, middle class, working class, working poor) in order to maintain it, and the government helps with this role, in part by providing a public education that sets the population apart into places within the social class hierarchy. As you read, think

DOI: 10.4324/9781003322566-5

about what role education, and mathematics education, might play in maintaining these social relations reflected in the economic order. Because some may be confronted with these critiques for the first time, it's important to note early on that understanding these critiques does not require you to completely reject capitalism. We'll return to this point a bit later.

The overarching question to think about in reading our world through these texts is the notion of social class. In the United States, for example, study after study has reported the likelihood of one's income to reflect that of her family of origin. This points to the notion of social reproduction and a lack of social mobility within the economic scheme. Social class thus includes one's income level, and it also refers to occupational prestige. Research in the United States indicates the following classes: executive elite (corporate CEOs, VPs, and board members) as having the most prestige, followed by the professional class (doctors, lawyers, etc.), next the middle class (teachers, nurses, owners of small shops, middle management, and skilled laborers), and, finally, working class (unskilled labor, service sector employees, and factory employees). I borrow these categories from Jean Anyon's (1980) article "Social Class and the Hidden Curriculum of Work," which investigates schools and social class. We will return to this article in detail in the next section, but for now it is helpful in furthering our understanding of social class as a social construct.

Anyon suggests three factors that describe one's place in the economic order of a capitalist society. The first, "ownership relations," points to the distinction among classes regarding the wealth that is owned. For example, Anyon points out that persons in the executive class own the majority of stocks in the United States. A recent Pew Research study indicates that the wealthiest 5% in the United States own 62% of stock. The poorest 60% own 4.2% of stock in the United States. The second characteristic determinant of social class, "relationship between people," suggests one's level of status in the occupational workforce. Think of the differing levels of authority between boss and employee and how these correspond to the class levels outlined in the previous paragraph. Those in the higher classes have more authority over others and autonomy in their work. The third factor determining social class is "relationship between people and their work." Think of the factory worker on an assembly line who has little understanding of the role his specific part plays in the grand scheme of the product being created as compared to the executive in charge of the factory operation who determines the entire process and knows its parts in total.

The construct of social class is embedded within the capitalist economic order and it is important to understanding this context more fully. To do this we begin with Karl Marx's famous critique of capitalism. He first published *Capital, Volume 1* in 1867. As a critique of capitalism, he works within the logics of capital to demonstrate its inherent inequality. In other words, Marx does not discuss the potential for corruption or excessive greed, but describes how capitalism is structured to create inequality and necessitates the creation of class hierarchy.

First is Marx's point about the exploitation of labor in the market. He poses the following question: How is it that a capitalist can make a profit when he functions in a free market of equivalent exchanges? In the open market, it is presumed that goods and services are exchanged for materials at a price deemed fair to the two actors in the exchange. So, how does profit turn up suddenly in the hands of the capitalist? Marx proposes that one value in the market serves two roles, and that is the labor value. On the one hand, a laborer receives money for her work. Let's call this amount W. This amount is negotiated based on how much the laborer thinks she needs to earn to support herself. Assuming for the moment this is the clothing industry, let's say our laborer earns $10 per sweater she puts together out of yarn she assembles together. On the other hand, the laborer adds value to the products on which she is working; let's call this P. The capitalist buys the yarn needed for sweaters at $20 per sweater, but sells the sweaters for $100. This means that the laborer has added $80 of value, minus of course other costs that the capitalist has to pay such as the knitting needles, building where the knitters work, etc. So P is, let's say, approximately $50 per sweater. The laborer has added a value of $50 to the materials but is only paid $10. In this sketch, you see how P is greater than W and this, Marx argues, is the exploitative situation by which capitalist profit emerges. As you can see, such an argument rests on mathematical knowledge to understand how capitalism exploits labor.

In order for the capitalist system to survive, there needs to be a consistent supply of laborers upon which the capitalist class can draw. Marx argues that the system reproduces itself, giving the rise to class distinctions such as the capitalist class and working class:

> Capitalist production therefore reproduces in the course of its own process the separation between labour-power and the conditions of labour. It thereby reproduces and perpetuates the conditions under which the worker is exploited. It incessantly forces him to sell his labour-power in order to live, and enables the capitalist to purchase labour-power in order that he may enrich himself. It is no longer a mere accident that capitalist and worker confront each other in the market as buyer and seller. It is the alternating rhythm of the process itself which throws the worker back onto the market again and again as a seller of his labour-power and continually transforms his own product into a means by which another man can purchase him.
>
> *(Marx, 1990, p. 723)*

In the separation of the laborer from the means of production, the working class looks to the capitalist for means of survival. In most cases, the initiation of such relationships occurs alongside a surplus of people forced from their previous means of survival. For example, Marx describes the expropriation of peasants from their land in fifteenth-century England. As a result, ownerless individuals look to find paid employment to engage and survive within the market system they are thrust into.

Looking at over a century of capitalist economic structures, neo-Marxist David Harvey describes the balance that can occur between the working class and the capitalist class. First off, the struggle and triumph of collective bargaining presents the working class' best efforts at reigning in the exploits of the capitalist class. Here, workers form unions to negotiate wages collectively; doing so as a large group gives each laborer more strength in struggling against a capitalist whose interest is in exploiting labor. Many examples of unions show that collective bargaining effectively lowers capitalist rates of exploitation. Progressive nations have laws that support the creation and maintenance of unions. Harvey writes, "A 'class compromise' between capital and labor was generally advocated as the key guarantor of domestic peace and tranquility (p. 10)." However, Harvey suggests how such compromise over exploitation only occurs when capitalists enjoy economic prosperity.

When capital power weakens, as in the United States in the 1960s, capitalists target the class compromise. Harvey and others describe this, the modern era, as neoliberalism, in which the political and economic elite aim to draw back compromises between capitalists and labor. The neoliberal era assumes the logics of a capitalist-free market on all aspects of the ideal social life. In the capitalist-free market, all individuals are assumed to behave rationally in their own self-interest. A capitalist makes the best decisions with respect to hiring labor, purchasing raw materials, etc.; a laborer sells his labor to the capitalist who will compensate the best, etc. In this world, everyone competes for greater profits and rewards. Under neoliberalism, modern social life should assume such practices.

Many who espouse neoliberalism think that public schools, for example, should compete for students; such competition would engender in schools their pursuance of self-interest and, consequently, wise decision-making. As Harvey notes, the neoliberal era requires a state to simultaneously support the free market and get out of its way:

> Neoliberal is in the first instance a theory of political economic practices that proposes that human well-being can best be advanced by liberating individual entrepreneurial freedoms and skills within an institutional framework characterized by strong private property rights, free markets and free trade. The role of the state is to create and preserve an institutional framework appropriate to such practices. The state has to guarantee, for example, the quality and integrity of money. It must also set up those military, defense, police, and legal structures and functions required to secure private property rights and to guarantee, by force if need be, the proper functioning of markets. Furthermore, if markets do not exist (in areas such as land, water, education, health care, social security, or environmental pollution) then they must be created, by state action if necessary.

But beyond these tasks the state should not venture. State interventions in markets (once created) must be kept to a bare minimum because, according to the theory, the state cannot possibly possess enough information to second-guess market signals (prices) and because powerful interest groups will inevitably distort and bias state interventions (particularly in democracies) for their own benefit.

(Harvey, p. 2)

In the neoliberal era, public schools should function for the needs of the free market system.

You may be wondering if teaching with a critical perspective requires you to be anti-capitalist. This is certainly one option you can take, and I encourage you to learn about alternative economic structures that might take its place, such as anarchist theories of economics. The concept of anarcho-syndicalism is a good place to start, and you can read Noam Chomsky's introductions on the topic (see reference list at the conclusion of this chapter). Alternatively, you can argue for fairer class compromise within the existing structure of capitalism. Learning of the exploits inherent in capitalism positions you to examine the status quo critically and push for what changes can be made given the current conditions and circumstances. For this option, a sort of reining in of capitalism, we can turn to John Maynard Keynes and what is now known as Keynesian economics. This is similar to the balance between working and capitalist classes that we learned of earlier through David Harvey's work. In Keynes' 1936 *The General Theory of Employment, Interest and Money*, he argues that the free market cannot maximize employment for the working class. Left to its own devices, there will exist significant unemployment and poverty. This implies the need for government interventions and job creation. Contemporary economists promoting this view include Robert Reich and Joseph Stiglitz. Teaching for equity requires you to understand the exploitation inherent in capitalist logics and think through the range of possibilities, from alternative economic structures like anarchism and socialism to more progressive capitalism like Keynesian economics. In the next section, we think about how the capitalist construction of social class hierarchies relates to education.

Class reproduction, schools, and critical pedagogy

Reading about the exploits of capitalism and the lack of social mobility within the system may cause you to wonder why individuals and groups, especially those in lower social classes, do not fight to change it more often than they do. The answer to this question lies with a look at government provided education. Contrary to what neoliberal thinking will tell you, capitalism requires government structures for a few key components. The enforcement of a law on private property is a clear example. Without a government sponsored and enforced law of ownership, capitalists cannot own the means of production and thereby exploit laborers. Additionally, government-run public schools reproduce social class hierarchies necessary for the reproduction of social economic hierarchies. In this section, I will briefly review both theories on and observations from public schools, especially with reviews of contributions by Pierre Bourdieu and Jean Anyon. In addition, we will review a significant strand of thought regarding teaching against the reproduction of class hierarchy: the Marxist-inspired critical pedagogy.

Schools prevent mass uprising from the populace because they engender a feeling that we live in a meritocracy. Such an ideal describes a society where individuals in positions of power, authority, and wealth have achieved their status as a result of merits they have attained. In other words, working hard in school leads to success in life. Some of us proudly display our success as a school student in order to demonstrate our qualifications for a role in society, be it a managerial position, political office, etc. However, theorists who recognize the necessity of social class for the economic order question meritocracy, claiming that its mythology leads to self-blame for one's lack of achievement. Those of us not doing well in school, and subsequently attaining roles in society deprived of power and wealth, blame ourselves for inadequate brains and/or poor work ethic. To consider public education as a servant for capitalist society requires that we consider meritocracy as a myth: How does public education perpetuate and reproduce social class, rather than allow for social mobility? It turns out there are two related answers. First, school culture and its curriculum more closely match the culture and experience of upper- and upper middle-class life, thus putting students from higher social class backgrounds at an advantage. Second, schooling experiences are differentiated according to social class, with students in lower social classes receiving a schooling program that relates to lower social class work expectations, and, similarly, students from higher social class backgrounds receiving an education in what is required to work in higher-class careers.

For the first, we look to the French philosopher and social theorist Pierre Bourdieu. In a chapter titled "The Forms of Capital," he describes three concepts of capital: economic, cultural, and social. Economic are those assets that can be immediately converted into material wealth; cultural as aspects to an individual that can be converted into economic capital, for example in the form of knowledge from education; and social, or the networks of "who you know," which can be converted into economic capital. Of the three, cultural capital relates most to the educational process. Think of cultural capital as the knowledge, behavior, and habits of mind that reflect one's status in the economic order. Essentially social reproduction works through the schools via the notion of cultural capital because schools reflect the culture of the upper and upper middle class. Thus, a student from the upper and upper middle classes will learn this culture both at school and elsewhere in life, at home for example. The student has advantage in the school setting, whereas a working-class student does not. When it comes to measures of school performance, it is inevitable that the upper- and upper-middle-class students on the whole outperform the lower-class students. As Bourdieu suggests, the stratification of "succeed and not succeed" reproduces the social classes and limits social mobility.

For the second way that schools reproduce social class, we look to educational scholar Jean Anyon. Instead of merely providing everyone a similar education where upper-class students have advantage, Anyon documents that schools are differentiated by economic class and provide differing instruction. Her 1980 article "Social Class and the Hidden Curriculum of Work" describes her research in New Jersey schools that are stratified according to the following social classes: executive elite, affluent professional, middle class, and working class. All schools used the same textbooks and similar curriculum; however, her observations of the instruction demonstrated stark contrasts. In the working-class schools, work is procedural and textbook oriented, usually involves practice worksheets, and generally follows directives from an authoritarian teacher. Contrast that with the executive elite school, where work is characterized as developing analytical thinking, debating opinions on social issues, writing original essays, and having some freedom to pursue individual interests. Anyon also noted distinctions regarding the attitude with which teachers approach their students. She concludes that a "hidden curriculum" is at play, in which students learn their relationships to others as dictated by their place in the social economic order. For example, the obedience to authority required by factory workers is what is taught in working-class schools.

Having understood this critique of schools in the ways they reproduce social class hierarchy, as mathematics teachers working for greater equity we need to look at theories of practice that will interrupt such social reproduction. Critical pedagogy, which now comprises a host of subfields, emerged initially as a Marxist interpretation of the role teaching and learning can play in raising class consciousness, a central feature of Marx's call to action in disrupting social class hierarchy. Initial work in developing the concept is typically credited to Brazilian educator Paulo Freire.

Freire's *Pedagogy of the Oppressed* (2000, first published 1968) has taken on great meaning for teachers with a critical perspective. Essentially, Freire describes a naïve consciousness in which typical ideas are held about groups of people and the way society functions, and especially those ideas that keep power structures as they are. Working in Brazil and elsewhere in Latin America, Freire's practical work involved the oppressed, working class, which internalized a negative image given the social class relations set forth by the upper class. Freire led the practice of teaching literacy through codifications, or images depicting an individual's position in the political economic order. Images prompted discussion and eventual participation in the written word, all motivated by the learner's development of a critical consciousness that replaces the false naïve consciousness. In Portuguese, Freire gives the term *conscientizacao* to describe such an awareness of new consciousness.

The idea of raising consciousness has come to be relevant for each and every of the chapters of this book. The concept of teaching to raise consciousness, as above, can be applied to race and gender constructs just as it can be useful for teaching class consciousness. Given its consistent thinking with Marxist theory (false consciousness) and Freire's initial goals for Latin America, I introduce it here in the chapter on class and education. His pedagogy increased awareness of the oppressive relationship in the economic order and motivated revolutionary thinking about economics and social relations. Many have followed this tradition, including several scholars in the United States who were students of Paulo Freire. These include bell hooks, Peter MacLaren, Donaldo Macedo, and Antonia Darder.

This section highlighted major discussions on class and education. On the one hand, we have a great many studies and concepts that criticize education for its role in social reproduction of economic class, and on the other are important contributions from Freire and others that teach us how to interrupt production. Seeing these contributions in two camps, then, as mathematics educators we must ask ourselves the following questions: (1) What elements in my practice of mathematics education reinforce class structure and limit social mobility? (2) What efforts can I make to change these elements, or perhaps more explicitly teach to raise consciousness when teaching mathematics? The next section reviews important considerations on these two questions.

Social class and mathematics education

With broader understandings now at hand, this section digs deeper into the thinking on social class relations and mathematics education. Similar to the previous section, there are contributions regarding how mathematics education reproduces social class relations as well as efforts to directly interrupt reproduction. As for the reproduction of social classes, we need to think deeply about what school mathematics is. How does school mathematics reflect upper- and middle-class culture and thus give advantage to students from higher social classes? How are mathematics education programs differentiated according to social classes, where some students are offered a different version of teaching and learning than others? Finally, a host of direct applications of Freirian theory and practice to mathematics education will be discussed.

Researchers across the globe have produced empirical evidence for mathematics education's role in reproducing social class relations. The article "Structural exclusion through school mathematics: using Bourdieu to understand mathematics as a social practice" by Robyn Jorgensen, Peter Gates, and Vanessa Roper provides case studies documenting how mathematics education effectively includes and excludes students based on their social class backgrounds. The authors provide examples showing that students from the United Kingdom who come from higher-class homes have an advantage in learning school mathematics.

For example, in the early years parents and caregivers "play school" and ask "school-like" questions more readily than their lower-class counterparts. This does not imply that lower-class parents and caregivers are less able or smart to do so and instead highlight the cultural mismatch between their homes and school life and congruence between higher-class homes and school life. As students work through the schooling system, opportunities to engage in learning the culture of school differentiate through structures like tracking into advanced-, general-, and lower-level programs and are further reinforced by differences in advantage back at home. As shown by the case studies by Jorgensen et al., students from higher-class homes might have experiences with this school culture, the behaviors and habits of mind required by mathematics education, whereas lower-class students might not. A more recent finding from Quaye and Pomeroy (2021) indicates the stratified mathematics education experiences for students. They demonstrate that working-class students have a less-favorable relationship to mathematics, whereas middle-class students "reported more positive attitudes towards mathematics, more positive perceived parental attitudes towards mathematics, and had higher mathematics achievement" (p. 155).

Providing research in the US context, Sarah Theule Lubienski (2000) offers additional insights on social class with a specific discussion about reform math teaching. Recall the push for mathematics teaching and learning that emphasizes student-centered learning, discovery, problem-solving, manipulatives, etc., as discussed in Chapter 2, and the structure for lesson planning that requires students to experience an activity and then debrief in whole-class discussion. Lubienski's work requires that we reconsider these strategies carefully for how they might reinforce social class reproduction. Her article "A clash of social class cultures? Students' experiences in a discussion-intensive mathematics classroom" draws on interviews and observations from a reform mathematics classroom to conclude that some aspects, specifically the teacher prompts that lead to whole-class discussion, more readily match the homelife of students from higher-class backgrounds. Again, it's a match of the home culture and school mathematics for higher-class students and a mismatch for lower-class students. Lubienski's work describes how the open-ended, abstract thinking questions required by reform efforts were less familiar for lower-class students who instead preferred questions with answers to specific, contextualized problems. This does not suggest that lower-class students cannot do the higher-order thinking questions, but that the teacher must be aware of how she structures these opportunities and consistently reflect on whether students are being advantaged/disadvantaged in her classroom due to social class.

There are also a host of contributions in mathematics education that use critical pedagogy as a perspective to teach students in mathematics classrooms about social class relations. Many of these draw on the writings and practice of Paulo Freire, discussed in the earlier section. In the early 1980s, Marilyn Frankenstein applied critical pedagogy to mathematics education both theoretically and in practice, especially in her work teaching basic skills mathematics courses for adult learners. Her 1983 article "Critical mathematics education: An application of Paulo Freire's epistemology" is pointed to as a clear introduction to teaching mathematics that interrupts typical social class relations. She focuses on the problem-posing nature of Freire's approach to raising consciousness and suggests mathematics' role in social class relations and mathematical knowledge as key to understanding it. For example:

> A mathematically illiterate populace can be convinced that social welfare programs are responsible for their declining standard of living, because they will not research the numbers to uncover that 'welfare' to the rich dwarfs any meager subsidies given to the poor. For example, in 1975 the maximum payment to an Aid for Dependent Children family of four was $5,000 and the average tax loophole for each of the richest 160,000 taxpayers was $45,000.
>
> *(p. 327)*

Frankenstein's work catapulted efforts to teach mathematics for liberation in the Freirian tradition. Although many others can be included here, a comprehensive and celebrated study in these efforts is Eric Gutstein's (2006) *Reading and writing the world with mathematics*. Here he provides several lesson examples of teaching mathematics to interrupt social class hierarchies. These examples are richly detailed with documentation of urban, low-income student experiences with such a curriculum. It should also be noted that the work has a clear intersectional focus, attending to students' social class and BIPOC racial/ethnic identities. The book's countless examples inspire us for the successes we can have in teaching mathematics that raise consciousness. Gutstein's contribution details the intersections specifically of race and class and the role that typical mathematics education plays in this as well as the ways that an alternative mathematics education can push against it.

In concluding this section on mathematics education's relationship to social class, we can revisit the major considerations put forth by the research. First, we have to consider the ways that students from higher-class backgrounds are given advantage in mathematics education. This could be because homelife corresponds more readily to the mathematics education classroom in a number of ways as well as the differentiated teaching and learning experiences in mathematics that are afforded to students from differing social class backgrounds. Second, schools privilege mathematics instruction because its content and teaching/learning experiences correspond to social class relations. Mathematical content is necessary for science-based industries and the typical teaching/learning of mathematics reinforces obedience to authority. Finally, we have much to be inspired by with the applications of a Marxist pedagogy to mathematics instruction. In interrupting social class reproduction, we are called to a careful study of the teaching and learning modeled for us by Frankenstein and Gutstein.

Activities and prompts for discussion:

1. Use the frameworks for teaching and learning mathematics that were introduced in the second chapter and your preferred lesson plan format to create a mathematics lesson in the Freirian tradition. This lesson must teach both a content and a process standard suitable to your grade level and provide an experience through which students will begin to think critically about social class relations. Below, an example scenario is suggested if you are having trouble getting started:

   Consider writing a lesson that teaches students the Marxist concept of exploitation of labor (discussed in the first section of this chapter). To explore this notion, students will be required to employ both number and operations thinking and algebraic thinking. The way you set up the example problems and scenarios will depend on the grade level and student readiness. On the lower end, use more examples with numbers and have students compare and contrast whether one capitalist exploits more than another. On the higher end, employ algebraic thinking earlier in the lesson and require students to debate algebraically defined rules that curb a capitalist's rate of exploitation. As an example, students might come up with: the amount paid to the laborer cannot be less than 2.5 times the added value the laborer provides to the product. Give students multiple fictional examples of a capitalist's labor costs, material costs and means of production upkeep. Ultimately all students will have discussions about what is fair and some might question whether capitalists should own the means of production at all, or if perhaps all laborers should collectively own and share profits equally.

2. Discussion with peers: Reflect on your own schooling experience. Would you say that the elementary school you attended was homogenous or diverse with respect to social class? What kind of teaching and learning, generally and specific to math, took place there? Compare and contrast with your peers to determine the extent to which your experiences match Anyon's research on schools of differing social classes.

3. Work with a partner to interview two students about their perceptions of "math class." Ask questions related to their perceptions of what mathematics is and what is expected of them during instruction. You want to get the sense of whether students feel that they must obey rules or whether they are encouraged to think critically, communicate, and justify their mathematical reasoning. With your partner, compare and contrast the two interviews noting how they might correspond to the notions of social class related to this chapter. Do you think that your students' class backgrounds and cultural expectations at home are similar to or different from what is expected of them in the mathematics classroom?

Classroom tips:

- Interrupt social class reproduction by requiring all students to think critically and reason with mathematics. Avoid the tendency that we see in the research: students from lower-class backgrounds are only offered procedural mathematics.
- Incorporate contexts and lessons with economics to teach students about social class relations and the inequities that exist. This will raise student consciousness, preparing them better for their own individual success in the economic system. It also furthers their worldview and will likely increase their civic participation and activism toward greater economic equity in society.
- Consider how your classroom practices and expectations for students correspond to the social class backgrounds of your students. What are the areas where your students from higher-class backgrounds are given an advantage in your classroom? How can you support your students from lower-class backgrounds, holding them to the same goals for learning but providing them with supports to get there?

TABLE 5.1 Important terms and concepts in this chapter

Social class	The social construction of class is the reproduction of a hierarchy of economic groups and general lack of social mobility; social class levels are marked by one's own wealth, level of authority vs. obedience at work, and amount of autonomy at work; in the United States, typical levels include executive elite, affluent professional, middle class, working class, and working poor
Cultural capital	As the knowledge, behavior, and habits of mind for success in the economic order, this is what is taught and assessed by schools; upper-class students have advantage because they learn it both in and out of school.
Critical pedagogy	A theory-informed practice (praxis) in which teaching and learning emphasizes raising consciousness about social class relations; the concept has expanded and been applied to pedagogies of raising consciousness about many, if not all, forms of oppression beyond social class hierarchy.
Exploitation of labor	The central critique of capitalism in which a given laborer is not paid the true value of their contribution; to reduce exploitation of labor, collective bargaining (unions) strengthens the laborer's side
Neoliberalism	Advanced stage capitalism that seeks to reduce working-class power and accumulate more private ownership of public goods
Social reproduction	Social class hierarchies are reproduced generation after generation with limited social mobility for individuals (there are individual exceptions but the pattern clearly expresses lack of social mobility on the whole)

(*Continued*)

TABLE 5.1 (Continued)

Myth of meritocracy	Public education as a free and supposedly equal opportunity provides the myth that students have social ability; if they end up in a lower-income job, this is the result of their own work performance in schools; the reality is that schools are set up to reproduce social classes given their advantage and privilege to students with higher social class backgrounds
Keynesian economics	A capitalist approach to economics that seeks for more balance between working class and ownership (capitalist) class
Anarcho-syndicalism	A non-capitalist approach to economics in which workers collectively own the means of production and distribute value equitably

Further reading

Anyon, J. (1980). Social class and the hidden curriculum of work. *Journal of Education* *162*(1): 67–92.

Bourdieu, P. (1986). *The forms of capital.* Retrieved from https://www.marxists.org/reference/subject/philosophy/works/fr/bourdieu-forms-capital.htm

Chomsky, N. (2005). *Chomsky on anarchism.* AK Press.

Frankenstein, M. (1983). Critical mathematics education: An application of Paulo Freire's epistemology. *Journal of Education* *165*(4): 315–339.

Freire, P. (2000). *Pedagogy of the oppressed.* Bloomsbury Academic.

Gutstein, E. (2006). *Reading and writing the world with mathematics: Toward a pedagogy of social justice.* Routledge.

Harvey, D. (2005). *A brief history of neoliberalism.* Oxford University Press.

Jorgensen, R., Gates, P., & Roper, V. (2013). Structural exclusion through school mathematics: Using Bourdieu to understand mathematics as a social practice. *Educational Studies in Mathematics 87*: 221–239.

Keynes, J.M. (2009). *The general theory of employment, interest and money.* Classic Books America.

Lubienski, S. (2000). A clash of social class cultures? Students' experiences in a discussion-intensive seventh-grade mathematics classroom. *The Elementary School Journal 100*(4): 377–403.

Quaye, J., & Pomeroy, D. (2021). Social class inequalities in attitudes towards mathematics and achievement in mathematics cross generations: A quantitative Bourdieusian analysis. *Educational Studies in Mathematics 109*: 155–175.

Marx, K. (1990). *Capital Volume 1.* Penguin.

6

GENDER TROUBLE

Rationalism vs. masculinity in mathematics education

Are there biological differences between men and women that require they be educated differently? Why do girls outperform boys in mathematics in the earlier grades but as adults boys are more likely to end up in mathematically intensive careers? How do mathematics teachers perceive students who are successful and not successful in these classes, and why do they give differing responses when discussing boys versus girls? These and other questions will be explored in this chapter, where we first trouble gender by affirming that these categories are social constructions rather than biologically based realities. Next we examine perspectives on gender to education in general with a look at important contributions from the literature before reviewing what mathematics education specifically says about gender and our teaching of math.

Here is a brief, somewhat recent news event to motivate our work in this chapter. Lawrence Summers is a world-renowned economist who in the late 1990s served as US president Bill Clinton's secretary of the treasury. Over the years, he had gained accolades for his work as an academic and in his service to banks and hedge fund management companies. In 2005, when he was the president of Harvard University, he made a statement at an economics conference that shook the mathematics and sciences world. Summers gave a speech highlighting the gender imbalance that greatly favors men over women in mathematics and science departments at colleges and universities. He pointed to a variety of possible explanations but suggested that he favors differences in cognitive ability over discrimination and cultural factors. To prove the point, Summers referenced difference in aptitude tests administered in the twelfth grade. Needless to say, his speech caused quite a stir and is possibly the reason why the following year Summers stepped down as Harvard's president. This story begs you to think through the validity of Summers' claim. The simplest critique, as we shall see, is

DOI: 10.4324/9781003322566-6

that the classification of people into "men" and "women" is far less straightforward than he makes it seem. This means that it becomes impossible to accurately document a distinction in aptitude according to gender. We'll start by taking this much-needed step back to think through what we mean by gender, beginning with this question in the first section of the chapter.

Gender and the history of feminism

To start, take the distinction between biological sex and gender. Biological sex refers to one's genetic makeup, physical anatomy, and hormones and includes the categories male, female, and intersex. We say that a sex is assigned at birth, for example, "Assigned Male at Birth." On the other hand, gender describes the typical behaviors, ways of being and other characteristics that are thought to be masculine and feminine. A transgender person feels and exhibits the characteristics of the gender that is opposite to their sex assigned at birth. The term "cisgender" describes people who express and feel the characteristics of the gender that was assigned at birth. There are also non-binary people who do not express and feel characteristics of either gender, or gender fluid individuals who may express and feel characteristics of both genders. There are additional gender identities beyond these as well.

Reviewing these examples helps to clarify that gender is something we perform, rather than a natural given fact. Advanced thinking on gender from the social sciences suggests that we are nurtured to perform our gender rather than born to behave a certain way. Not just about blue for boys and pink for girls, the characteristics associated with masculinity and femininity are taught to us early on and continuously throughout life by the family, media, schools, our workplaces, religion, etc. For example, masculine traits include competition, strength, rationalism, anger, courage, and assertiveness; feminine traits include dependence, emotionality, weakness, quietness, grace, and nurturing. As we grow up, we are encouraged, some would say forced, to perform the traits aligning with the biological sex we are assigned. For many whose biological sex does not align with their gender, this can be an extremely painful process. And taken in total, the societal expectation of such gender roles has significant consequences including the oppression and lack of opportunity for girls, women, and transgender individuals. We will be looking specifically at how this operates in the mathematics education space.

The previous paragraphs introduce the concept of gender as social construct, and I encourage you to explore these further by reading classic texts on the subject. Judith Butler is a gender and sexuality theorist with many books on the topic. Her book *Undoing gender* (2004) provides deeper theoretical considerations on several questions related to gender. For example, consider the following discussion regarding the intersex, or people who are born with ambiguous physical characteristics with respect to sex:

The question of surgical "correction" for intersexed children is one case in point. There the argument is made that children born with irregular primary sexual characteristics are to be "corrected" in order to fit in, feel more comfortable, achieve normality. Corrective surgery is sometimes performed with parental support and in the name of normalization, and the physical and psychic costs of the surgery have proven to be enormous for those persons who have been submitted, as it were, to the knife of the norm. The bodies produced through such a regulatory enforcement of gender are bodies in pain, bearing the marks of violence and suffering. Here the ideality of gendered morphology is quite literally incised in the flesh.

(p. 53)

With a discussion that "troubles gender" at hand, we now can think about how such constructions of femininity and masculinity relate to power structures throughout history. By ascribing gender roles, and their characteristics, to persons with particular physiologies, groups of people, specifically women and the transgendered, endure oppressive institutional structures and societal habits of mind positioning them as "less than" or inferior. Women are perceived as the inferior gender and transgender men and women have extremely marginalized gender identities. Although our discussion of gender here naturally included transgender identities, as this is typically discussed together with minoritized sexual identities, we will pick up transgender work again in the next chapter on LGBTQ+ and mathematics education. The remainder of this chapter primarily focuses on understanding women and girls in mathematics education, and we'll continue by looking next at the history of the feminist movement, which is typically described in waves.

First-wave feminism refers to the suffrage movements in several countries during the later nineteenth and early twentieth centuries. Second-wave feminism refers to the rebirth of gender equality discussions of the 1960s to 1980s, including a focus on sexuality, family life, inequality at work, and unequal educational opportunities. In the 1990s, a "third-wave" or "post-feminism" more strongly integrates queer and non-white perspectives in feminism as well as strong opposition to gender norms and binaries.

A concise and accessible review of the history of feminism comes from bell hooks, a Black feminist, in her text *Feminism is for Everybody*. She opens with some clear responses to the common misunderstanding of feminism, which she characterizes as follows:

I tend to hear all about the evil of feminism and the bad feminists: how "they" hate men; how "they" want to go against nature and god; how "they" are all lesbians; how "they" are taking all the jobs and making the world hard for white men, who do not stand a chance. When I ask these same folks about the feminist books or magazines they read, when I ask them about the feminist talks they have heard, about the feminist activists they know, they respond by letting me know that everything they know about feminism has come into their lives thirdhand, that they really have not come close enough to feminist movement to know what really happens, what it's really about. Mostly they think feminism is a bunch of angry women who want to be like men. They do not even think about feminism as being about rights – about women gaining equal rights.

(pp. vi–vii)

Whereas "feminism" is so often misunderstood, hooks reminds us that the term simply rejects that men are superior to women. This is not altogether different from an anti-racist stance rejecting that whites are superior to people of color.

One feature of hooks' review on the history of feminism is her attention to the emergence of feminism within and across other avenues of social justice. For example, in the civil rights movement and in movements for social class equity, hooks notes that women highlighted the expectations that they follow male leaders in these social movements. The targeted messages and actions of such movements for equality provoked inconsistent thoughts and feelings for the women who took up such causes. The problem was with "men who were telling the world about the importance of freedom while subordinating the women in their ranks" (p. 2). As hooks describes the history, early on feminists polarized among those that sought more sweeping social reforms aimed at identifying consistencies among oppression (racism, classism, sexism) and eradicating these together, and those that targeted the women's cause specifically with goals for equality in schools and the workplace. The latter had greater success, especially with the corporate powers that be and emerging conservative responses to the civil rights era already in place. These less radical feminist efforts, as hooks argues, contributed to the stalled progress of the civil rights movement:

We can never forget that white women began to assert their need for freedom after civil rights, just at the point when racial discrimination was ending and black people, especially black males, might have attained equality in the workforce with white men.

(p. 4)

She notes the embrace of the less radical version of feminism by most women and the location of more radical feminist thinking as restricted to academic circles. Referring to the more mainstream version as "lifestyle feminism," hooks notes that such conversation and action lack their political origins. The central aim of *Feminism is for Everybody* is to regain this, especially by highlighting the interplay between sexism and related oppression, including race and social class. As we shall see, a host of approaches exist in the work on gender and mathematics education and reflect a similar range of stances.

Recall the opening discussion regarding the social construction of gender. These conversations emerge with more recent theoretical work in feminism that seeks to displace the feminine subject at its center. Third-wave feminism, as a poststructural feminism, calls into question the concept of gender as Judith Butler clearly does in *Undoing gender*. As well, third-wave feminism provides direct responses to hooks' concerns with first and second wave. Third-wave feminism highlights the intersection of sexism with, for example, racism, given the writings of Black feminists, such as hooks herself.

Gender and education

With a discussion of gender and feminism at hand, we turn this onto education and schools in general with a review of a handful of important sources in the field. This will properly situate our review of what work has been done with respect to mathematics education and gender and how teaching mathematics with a critical perspective can more properly reflect advanced understandings of gender and sexuality. The first of these do not fully position us to teach as critically as we can, but nevertheless provide an initial departure point for moving us in this direction.

One seminal text that certainly got several conversations going is *Women's ways of knowing: The development of self, voice, and mind* by Mary Field Belenky, Blyth McVicker Clinchy, Nancy Rule Goldberger, and Jill Mattuck Tarule (1986). The authors summarize their research on 135 women from varying social class, educational, and race/ethnicity backgrounds to describe five developmental positions characterizing the ways that a woman can understand, interpret, apply, and critique knowledge. The categories run from a "silenced knowledge" to a "constructed knowledge" and characterize the knower's relationship to the knowledge. This includes how she was taught knowledge, what she finds as more and valuable forms of knowledge, and whether she sees ideas as static or dependent on context. For example, the third developmental position is titled "subjective knowledge," where the learner has distrust for analytical ways of knowing in favor of more concrete thought processes rooted in experience. To contrast, the highest position, "constructed knowledge," considers all forms of knowledge including both analytic and experience-driven ideas primarily with the clear acknowledgment that an individual's knowledge is constructed by her experiences and analysis. For example, the constructivist knower listens to all perspectives and would begin a respond to a question with "My understanding is…"

Primarily, Belenky et al. provide a dialogue in response to theories of knowledge that up until then were dominated by male-centric frameworks. As we will see in the next section, *Women's ways of knowing* have been directly applied to mathematics education as well as several other disciplines. As well, the work has been perceived more deeply by third-wave feminists who take up the poststructural standpoints introduced in the previous section. Although

Women's ways responded directly and necessarily to a male-dominated viewpoint on psychology, it reinforces the gender binaries that the third wave aims to desta-bilize by perceiving women in a unitary manner. Thus, we need to continue our study by looking to the work of poststructural feminist applications to education.

Carmen Luke and Jennifer Gore situate their collection of scholarly writings on feminist pedagogy (*Feminisms and critical pedagogy*, 1992) within poststructural feminism: "That is, texts, classrooms, and identities are read as discursive inscrip-tions on material bodies/subjectivities. Pedagogical encounters and pedagogical texts are read both as a politics of signification and as historically contingent cultural practice" (p. 4). The collection of scholars in this work particularly think through critical pedagogy (discussed in detail in Chapter 5) as a "regime of truth" itself containing particular discourses that fail to fully challenge gender binaries and hierarchies.

> Hence, our poststructuralist feminist task is to go beyond the deconstruc-tion of the normative masculine subject valorized as the benchmark against which all others are measured, and to examine how and where the femi-nine is positioned in contemporary emancipatory discourses (including feminist discourses). The high visibility of "gender" in social justice and equity programs and policies, and its status in almost all progressive peda-gogical tracts, easily obscures the fact that equal space and representation in curriculum, policy, or the conference agenda does not in itself necessarily alter the status of the feminine as an add-on category or compensatory gesture. As such, the poststructuralist feminist agenda remains focused on challenging incorporation and marginalization, even and especially in lib-eral progressive discourses that make vocal claims to social justice on behalf of marginalized groups while denying their own technologies of power.
>
> *(pp. 6–7)*

When considering gender and educational reform on a deeper level, then, we need to think through the relations of power expressed by these reforms, considering to what extent they are reflective of simple adjustments to superfi-cially include the minoritized or whether these measures truly reflect significant change to practice and social relations by interrupting gender norms. In addition to a thoughtful, feminist reconsideration of critical pedagogy, the same must be done for a host of educational reforms in which gender is prioritized. In the next section of this chapter, we will look at a book containing critical feminist reviews of said reforms specific to mathematics education.

For now, we continue with some specific examples of how a feminist critique can modify our thinking on emancipatory education. One of the authors con-tributing to *Feminisms and critical pedagogy* is Valerie Walkerdine, a feminist psy-chologist who contributed significantly to gender and education and gender and mathematics education. In her chapter from the Luke and Gore text, "Progressive Pedagogy and Political Struggle," Walkerdine reflects on her primary school teaching of the 1960s, full of "progressive" intentions to liberate poor, inner-city children from the chains of oppressive economic and state structures. Upon this reflection, she considers the accordance such a project has with *maintaining* the oppressive structures rather than eradicating them. One of her primary examples focuses on the reductive, gendered role of the teacher in such a progressive pedagogy. The teacher, as a political entity, a "bourgeois and nurturant mother" (p. 20), aims to reject any power that might exist in her authoritative presence and through the freedom she provides will transform students into rational thinkers. However, the denial of the power that exists within the schooling structure presents a paradox for the teacher whereby students are not free, social class hierarchies are reproduced, and the teacher reduced to a feminine, passive subject.

Walkerdine's and other chapters in the anthology also point to the preference for rationalism within critical pedagogy as a masculine priority that reinforces surveillance and controlled societies. This bears significant relevance to the teaching and learning of mathematics, a discipline that can play a big part in categorizing and managing populations. These and the other critiques contained in the volume sit within the third wave of feminism that more carefully draws connections between forms of oppression (social class and gender, for example). This is especially due to the poststructural nature of critique in which the discursive messaging contained in social projects are laid out and scrutinized for their reproduction of particular oppressive structures. As we turn to mathematics education, consider the following exercise. Think about a recent reform in the teaching and learning of mathematics, and perhaps specifically one that aims to increase success by female students. How has this reform addressed gender and mathematics, or, in what ways is this reform reinforcing gender binaries? The answer must begin by identifying the problems with respect to gender and mathematics, to which we now turn in the next section.

Gender and mathematics education

Possibly more so than the other social identities discussed in this book, research on gender and mathematics education is plentiful. Several researchers have synthesized this work into a variety of categories that reflect the varying factions of feminism and those that we will look at here claim that some of the researchers, more than others, make significant progress toward gender equity. When examining studies that focus on gender and mathematics education, here are a few questions to consider: To what extent does the research view women and girls as lacking something needed to learn mathematics and/or having a pathology that prevents them from doing so? This would be a concerning approach and research to avoid. On the other hand, do the researchers imply a shift in the culture and practice of mathematics education or an adaptation that teachers and caregivers must make to facilitate greater success by girls? Do the researchers define gender in their study, and, if so, does the definition relate to a poststructural emphasis on the historically constructed nature of gender binaries?

We begin with a synthesis of the literature on gender and mathematics education from Rogers and Kaiser's (1995) edited book *Equity in mathematics education: Influences of feminism and culture.* In the introductory chapter, the editors lay out a helpful trajectory that summarizes the evolution of research on this topic. Using a framework developed by the feminist Peggy McIntosh, Rogers, and Kaiser present the phases of mathematics education and gender research as follows:

> Phase One: Womanless mathematics;
> Phase Two: Women in mathematics;
> Phase Three: Women as a problem in mathematics;
> Phase Four: Women as central to mathematics; and
> Phase Five: Mathematics reconstructed.
>
> *(p. 3)*

Speaking broadly, the shift in phases starts from the masculine culture and majority-male status of mathematics and ends with changing the nature of mathematics and the culture of its practice. In phase one, a definition of the situation is made clear: mathematics is a male-dominated space. This resonates with our discussion of mathematics education as a white institutional space discussed in Chapter 4. Despite the odds against it, some women attain status in the space (as research mathematicians, for example) and reflections shared by these individuals indicate the "silence and exclusion (p. 4)" they experience. In phase two, this problem is addressed superficially with token celebrations of the few women who make it within the male-dominated mathematical world. In response to the problem, the remaining phases make attempts to understand *why* fewer women achieve high status in mathematics. Phase three locates the source of the problem on the women themselves. The phase includes research on and promotion of intervention programs aiming to increase girls' interest and success in mathematics, something that certainly continues to this day and has transformed into STEM education camps and other opportunities for girls. These interventions aim to fix a situation where girls do not seem interested in mathematics, thereby locating the problem within them. In this case, girls and women are said to have a deficiency. Something they lack is a desire to learn mathematics unlike their male counterparts. As Rogers and Kaiser point out, another strand of this approach is the study and theorization of math anxiety as a source of the problem. This is an example of an effort to pathologize girls and women; in this case, girls and women have an illness that their male counterparts do not. In all cases, these attempts locate the problem within girls and do not work to change the system, but work within it (p. 6).

To this point, phase four addresses the problem with a shift toward the system. This research focuses on girls' and women's experiences in learning mathematics and takes a cue from feminist work in psychology, reminding us that how people learn might be influenced by the social identities by which we are positioned.

As Rogers and Kaiser put it, phase five suggests that mathematics will leave us with a new conception of mathematics, and in looking at a broader set of the work in feminist pedagogy and mathematics education, we see that this phase will include pushes beyond the theoretical considerations of phase four as well. This active, current phase can include the most advanced conception of gender, via poststructuralist and third-wave feminism, to address the intersectionalities of social identity at play as well as consider how conceptions of feminist pedagogy will interrupt the logics of women as an inferior Other. This thinking responds to the earlier stances resting on "the ways women think differently," something that can reproduce binary modes of thinking and perhaps perpetuate stereotypes. For the remainder of this section, we will focus in more depth on a few studies from phases four and five. Most importantly, these contributions no longer blame women for their lack of success in mathematics but look to the cultural practices of school mathematics that exclude and perpetuate gender stereotypes.

Recall the earlier section when we discussed the book *Women's ways of knowing* (Belenky et al., 1986), which problematized the male-centered theories about how people learn. Rossi Becker (1995) applies this directly to mathematics education and, to begin, provides the following caution in so doing:

> When I refer to women or girls in this chapter, as when the book refers to women in *Women's ways of knowing*, I am not automatically referring to all females. Here I am generalizing, not as mathematicians do, meaning every woman, but as social scientists do, meaning most women. The word 'women' is used to refer to all those individuals who think, come to know, or react in a fashion that is common to the majority of women. These individuals may be females or males. Also, the use of the word 'women' to describe the way some people think does not preclude the possibility that some women do not think in this way. There is, of course, the danger that acknowledging women's different ways of knowing will serve to reinforce stereotypes that demean women's capabilities. We have to make the case that 'different' does not mean one way of thinking is better than another. On the other hand, research can help provide evidence to support or refute the possibility of a women's way of knowing in mathematics and, if it is supported, demonstrate how this can help us understand and improve women's participation in mathematics.
>
> *(p. 164)*

These considerations on the limitations of generalizing and potential harm in reinforcing stereotypes are important cautions we must take when thinking about research on gender and mathematics education. Furthermore, this thinking is adequately addressed by the research rooted by poststructural feminist thinking which we will look at later in the section. Nevertheless, the application of

Women's ways of knowing is essential to thinking through gender and mathematics, and we now take a look at some of Rossi Becker's major themes.

Rossi Becker points to *Women's ways of knowing's* suggestion that a "connected teaching" is required to move women toward the higher levels of understanding, such as constructed knowledge. She applies the concept to mathematics:

> In connected mathematics teaching, one would share the process of solving problems with students, not just the finished product or proof. Students need to see all the crumpled papers we put in the wastepaper basket, if they are to understand that mathematicians do not arrive at a solution the first time or the first way.
>
> *(p. 168)*

The suggestions made to foster constructed knowledge resonate with many of the reforms of math education discussed in the second chapter of this book. She emphasizes how we must teach mathematics as a process, rather than a "universal truth handed down," that discovery methods should be in favor of traditional modeling and practice, and that multiple methods for problem-solving should be encourage in students. All of these are argued to increase success for female students and are consistent with the mathematics education reforms occurring within the trajectory of NCTM standards-based mathematics instruction.

Chapter 5's discussion of social class brought to light an interesting study that pushed back on the reform methods that are promoted by reform mathematics teaching. I bring up Lubienski's study again because it reminds us of the caution that we must take when considering the generalizations presented by *Women's ways*. To be sure, both the original study and its application aim to study women across social identities (race and class) and yet also articulate a generalized claim that most women think a certain way and will move toward this with a certain form of education. However, Lubienski draws on the literature of reform mathematics education to use whole-class discussions in enhancing mathematical instruction. In studying the girls in the classroom, she found that such efforts (used to promote connected thinking) actually silenced girls from lower social class backgrounds. These inconsistencies are exactly the pitfalls that poststructuralist feminist thinking takes on directly. It is important that we think through the multiple positions of individual learners rather than paint with broad strokes and make sweeping modifications that might lead us to consider our job complete. Such efforts that emphasize multiple social identities and reduce binary thinking reflect more advanced contributions in the conversation about gender and mathematics education.

Another chapter from the *Equity in mathematics education* edited book by Australian mathematics education researcher Sue Willis is titled "Gender reform through school mathematics" (1995). In line with the editors "Phase Four," this chapter criticizes research on gender and mathematics that locates deficits or pathologies in girls and supports the research that looks at the system of mathematics education and its specific exclusionary practices. Willis cites research (this is the work of Valerie Walkerdine, a researcher discussed earlier and to whom we will turn to again shortly) describing the culture of school mathematics that favors rule-following, whereas a challenge of rules and creativity is at the root of superior mathematical talent. This "sets up a conflict between what is regarded as necessary to achieve femininity and be a 'good girl' by doing what the teacher asks, and what is necessary to be regarded as 'really good' at mathematics" (p. 191). Similarly, Willis' chapter discusses several projects that implement novel curricula aiming directly to reduce gender bias and promote greater inclusivity. Interesting, mixed results are discussed, such as the "counter-sexist" mathematics textbook that due to its deliberate non-conformity *and* lack of comment resulted in students and teachers making jokes about the book rather than taking the learning of gender more seriously. As a result, Willis suggests that such intentions to disrupt gender bias need to be made clearer by teachers, authors, and researchers. She also highlights research where students with privilege in mathematics (typically male students) have openly resisted efforts for greater inclusion. However, had these students known that the efforts were to reduce gender bias in school mathematics, they may have been more receptive to novel interventions.

Moving more into phase five, we look at Valerie Walkerdine's important text *Counting girls out* (1998). First published in 1989, the text and its research are positioned firmly within poststructuralist feminism. Walkerdine states that her work rests on theoretical developments that put

> into historical perspective the construction of scientific ideas (or truths) about girls and boys, men and women, minds and Mathematics. It allows us to take apart these truths and their forming and informing of practices in which girls and women are taken to be poor at Mathematics.
>
> *(p. 18)*

Further,

> Our starting point is that there is no simple category 'woman' which can be revealed by feminist research, but that as feminists we can examine how facts, fictions and fantasies have been constituted and how these have affected the ways in which we have been positioned, understood and led to understand ourselves. Hence, while much feminist counter-research has attempted empirically to disprove the facts about girls' and women's performance, we felt that a fundamental problem remained. Accepting the categories and terms within which the issues were framed left feminist work always on the defensive and trapped within empiricism.
>
> *(p. 21)*

In the book, Walkerdine describes the fundamental character to school mathematics as the teaching of reason and the supremacy of rationalism. In her efforts to think of mathematical culture as historically embedded, she traces the roots of Cartesian rationalism to the enlightenment and highlights its gendered, specifically male-dominated, nature. Women were excluded and deemed irrational, their emotional characters and bodies only in service to the reproduction of reasoned man. In our classrooms, what students learn is not their facility with mathematics but their capacity to reason, to break free of emotion. This is the culture, practice, and ethos of the space of mathematics education. In addition, there exist steadfast gender binaries in our society, habits of mind that dictate who we are and how we should behave. Given that teachers and students both are subjected to constant messaging about these expected gender roles, it comes as no surprise, then, that girls and women are seen as outsiders to the space of mathematics education.

Walkerdine's extensive study of the mathematics education space supports this. The book recounts several studies across grade levels. One research project interviewed students and teachers to come to understand just what makes a student labeled "good" or "bad" in school mathematics. Walkerdine and her research team found stark contrasts. The following excerpt provides an example of this:

> Angela, then, is positioned classically as a 'good girl,' an 'ideal pupil,' but she does not have that elusive gift, 'brilliance.' She must rely on hard-work. But what has happened here? We saw how 'sensitive' Angela's mother was when the child was 4. She is articulate, one of the few children who found it easy to chat to her nursery school teachers. Why then at 10, even though she is outstanding in her class (surely coming top is an indication of achievement), is she designated a quiet, shy girl who comes top only through sheer hard-work? This judgment was never applied to any boy in any study—on the contrary, it was difficult for a boy to be judged a failure, even with the most appalling performance. Teachers talk, for example, about boys with very poor attainment still having potential or being bright.
>
> *(p. 88)*

The story of Angela is not unique and represents the consistent efforts Walkerdine and her research team took to inquire and discuss the school math experiences for girls. In study after study, Walkerdine finds teacher talk to indicate that boys who do well are gifted yet girls who do well work hard.

Since Walkerdine's seminal text, two books have continued this trajectory of poststructural feminist inquiry into mathematics education. *Masculinities in mathematics* by Healther Mendick (2006) continues the work of thinking through what characterizes good and bad math students. Subjects in her studies demonstrate that the gendering of school mathematics is just as much about social norms and expectations for work life as it is about "the gendering of maths itself, the gendered phantasies of rationality and of genius, and the ways these work with the construction of masculine and feminine as oppositional categories" (p. 67). A text clearly situated within third-wave feminism, Mendick discusses the "queering of mathematics education" as well. Along the lines of a poststructural refusal of binaries, one example of such efforts requires math teachers and students to reclaim the notion of ability. Throughout her research, exceptional mathematics students are thought to be born that way, yet we now know that who is really perceived as and encouraged to be rational falls along race, class, gender, sexuality, and other lines. A consistent rejection of the inborn nature of ability in mathematics education is an excellent suggestion for mathematics teachers to consider.

The second book by Sara Hottinger (2016) also carries on the tradition of poststructural feminist investigations of mathematics education. *Inventing the mathematics: Gender, race, and our cultural understanding of mathematics* requires that "We as a culture need to challenge, in a very deep and complex way, how we construct what it means to reason, what it means to think logically, and what it means to think mathematically" (p. 24). Hottinger provides exceptionally interesting poststructural analyses of mathematics textbooks, the field of ethnomathematics, and even the actress-turned female mathematician Danica McKellar. All of these endeavors relate to the objective of calling into question what and who mathematicians are and, as the title suggests, motivate a new conception for them.

With these deep inquiries into mathematics education and gender, we see clear habits of mind for math teachers to consider when teaching with gender equity. A starting point is recognizing and challenging assumptions about gender norms and our tendency to attribute a strong girl or young woman's performances to hard work rather than her brilliance. There are also gender equity practices that we can enact on a routine basis such as confronting the gender biases in story problems as noted by Rubel (2016). Although she notes that in the last 40 years "mathematics textbooks have been updated to reflect women and girls or men and boys in more diverse ways" (p. 437), teachers and students may still tend to stereotype with gender norms as they write mathematical scenarios for the classroom. We will return to Rubel's contribution again in the chapter on sexuality, as she points additionally to a related and worse problem with respect to sexuality stereotypes and exclusion of the LGBTQ+ community in mathematics education.

Activities and prompts for discussion:

1. Have an open and honest conversation with a fellow mathematics teacher or two with a focus on "our best mathematics students from the last few years." Without thinking too hard about the contents of this chapter, take time to describe the qualities that made them strong and successful. After you talk

about the students, reflect on your conversation together using what we discussed in this chapter. Did your descriptions of successful students fall into a typical gendered pattern, with boys as "naturally talented" and girls as "hard-working?" Discuss how you might try to go about interrupting these habits of mind in the future as you work with students.

2. Locate a section of a mathematics textbook that you use with students. Do you see any evidence of gender bias in the story or context problems included in this section? Draft five original story or word problems that actively interrupt gender norms.

Classroom tips:

- Represent gender equity in your classroom environment by including posters and biographical information about women and women of color mathematicians. When designing story and word problems, make sure to think carefully about gender stereotypes and author scenarios that interrupt typical gender norms. Discuss these changes openly with your students, making explicit your intentions to develop a gender inclusive space for your mathematics instruction.
- Avoid grouping students as "boys and girls" or addressing them as "ladies and gentleman." This reinforces a binary and fixed perspective on gender which is outdated and will not be an inclusive practice for all students. Address your entire class as "Students" or "Y'all" instead of "Guys."
- Reflect on your students' gender identities carefully, thinking about how you perceive their abilities and checking your own subconscious habits of mind that might reinforce gender norms. For example, have you caught yourself thinking that a female student "is strong because she practices math" but that a male student "is very smart but just needs to work harder?"

TABLE 6.1 Important terms and concepts in this chapter

Gender binary	A social construct of male and female that is typically confused with biological sex or what an individual is assigned at birth; individuals are expected to accord with stereotyped norms related to their gender; a false binary with several individuals identifying as non-binary
First-wave feminism	Initiated equality for women through suffrage
Second-wave feminism	Prioritized equality in home and workplace and other aspects of lifestyle
Poststructural feminism	Part of third wave of feminism, theory to undo gender norms and the gender binary; highlights the intersections of gender and other social identities including race, class, and sexuality
Cisgender	An individual whose sex assigned at birth corresponds to their gender expression

(Continued)

TABLE 6.1 (Continued)

Transgender	An individual whose sex assigned at birth is different from their gender expression
Masculinity	Gender norms and stereotypes typically ascribed to and expected of male individuals, including competition, strength, rationalism, anger, courage, and assertiveness
Femininity	Gender norms and stereotypes typically ascribed to and expected of female individuals including dependence, emotionality, weakness, quietness, grace, and nurturing

Further reading

Belenky, M.F., Clinchy, B.M., Goldberger, N.R., & Tarule, J.M. (1986). *Women's ways of knowing: The development of self, voice, and mind.* Basic Books.

Butler, J. (2004). *Undoing gender.* Routledge.

Foucault, M. (1990). *The history of sexuality: An introduction, Volume 1.* Vintage Books.

hooks, b. (2014). *Feminism is for everybody: Passionate politics.* Routledge.

Hottinger, S. (2016). *Inventing the mathematians: Gender, race, and our cultural understanding of mathematics.* SUNY Press.

Luke, C., & Gore, J. (1992). *Feminisms and critical pedagogy.* Routledge.

Mendick, H. (2006). *Masculinities in mathematics.* Open University Press.

Rogers, P., & Kaiser, G. (1995). *Equity in mathematics education: Influence of feminism and culture.* Falmer Press.

Rossi Becker, J. (1995). Women's ways of knowing mathematics. In G. Kaiser (Ed.) *Equity in mathematics education: Influences of feminism and culture,* pp. 164–174. Falmer.

Rubel, L. (2016). Speaking up and speaking out about gender in mathmathics. *Mathematics Teacher 109*(6): 435–439.

Walkerdine, V. (1998). *Counting girls out: Girls and mathematics.* Falmer Press.

7

LGBTQ+ WORK

Outing mathematics for heteronormativity and homophobia

Sexuality and gender expression, including trans and non-binary identities, deserve much attention in our classrooms even though this is an under-researched area in the mathematics education community. This chapter first draws on social theories about the social construction of sexual orientation, a review of research about the LGBTQ+ community, and what is known about LGBTQ+ students in general education scholarship. Applications of these concepts are directly related to the mathematics classroom, and examples are provided to indicate how a status quo mathematics classroom will reinforce both heteronormativity and homophobia. Practical examples of how to interrupt this culture are provided for mathematics teachers, including how to increase representation of the LGBTQ+ community into mathematics curriculum and language practices that teachers can model for their students.

Like gender, sexuality is based more on social norms and expectations than on biology. We refer to human sexuality and all its identities as social constructions, like we did with race, social class, and gender thus far. The language practices around human sexuality and identities are varied and unique to each individual and community; it is important as with all facets of human diversity that mathematics teachers stay up-to-date on current language practices. For one, it's difficult to pin down a unifying term to describe the broad range of identities that we are focusing on in this chapter; some prefer to use a simple but incomplete term, "queer," whereas others use an acronym like LGBTQ+. The latter acronym includes Lesbian, Gay, Bisexual, Transgender, and Queer, and the plus signifies other identities including Intersex, Asexual, and other identities. Christopher Dubbs, an expert in mathematics education and sexuality, describes the term "queer" as follows:

DOI: 10.4324/9781003322566-7

Historically, queer was linked with insult and shame. Today, queer has become the 'rallying point' not only for young gays and lesbians concerned with the homonormative images of gay men and lesbian women, but also for those whom wish to identify themselves with the anti-homophobic movement... Queer is understood as a false unifying umbrella, useful for solidarity with the necessary error of homogenizing and erasing differences to yield a temporary totalized identity, which then necessarily fails to represent the person. [I will use] queer as an umbrella term for lesbian, gay, bisexual, transgender, and other gender and sexual minorities.

(p. 1042)

As you engage with the literature on this subject, you will notice that varying authors use either an acronym or the term "queer" to signify the community. This chapter will use both LGBTQ+ and queer interchangeably to reflect the literature we will review. On these and more specific queer identities, there continues to be active debate on the best uses for language practices, as there are in most areas of human diversity, and we should be respectful when working with individuals, listening to their preference as to how they signify their own sexual or gender identity.

The focus in this chapter is on gender and sexuality minoritized identities. We briefly discussed marginalized genders in the previous chapter with clarification on transgender and non-binary identities. Although these groups and individuals face particular challenges in society and mathematics classrooms that relate more to gender than sexuality, because the queer and LGBTQ+ communities tend to group sexuality and gender minoritized identities together, we will do that as well in this chapter. In terms of sexuality and sexual identity, we are mostly referring to the varieties of romantic and sexual attraction that a person can have or not have for other people. Broadly speaking, the socially acceptable and more prevalent identity is a heterosexual sexual identity. Minoritized groups and individuals include same-gender and multiple-gender attractions as well as those with no attractions. A complete list of possible identities is impossible to determine but minoritized sexual identities include those identifying as homosexual, lesbian, gay, bisexual, asexual, homoromantic, biromantic, and aromantic. Sexual and gender identity is understood to be independent from each other and also fluid for individuals, meaning that it can change over the course of one's life. At the same time, many people feel they have a fixed gender or sexual identity that is the same for their entire life. Even though we say that our gender and sexual identities can be fluid, it is important to remember that a gender or sexual identity is not a "choice" made by an individual. Finally, as with all identities, intersectionality causes specific experiences for LGBTQ+ individuals that layer on to create unique experiences at the individual level.

Societal acceptance and norming of heterosexuality, as the dominant sexual identity, is pervasive throughout interpersonal interactions, media, and our institutions like religions, schools, and governments. This is called heteronormativity. Constantly, we deem minoritized sexual identities as "Other" by declaring preference and acceptance of heterosexual and heteroromantic life. When we watch a movie or a TV show, the romantic and sexual interaction will be straight. When an adult talks to a six-year-old boy, they will ask him what girls he has a crush on, assuming a sexual identity on a child. Similar expectations for gender normativity abound, just think of preferences for extracurricular activities like sports or art or music in the elementary school years. Homophobia, biphobia, and transphobia are prejudicial thoughts and discriminatory actions against minoritized sexual and gender identities. As with other facets of human diversity, discriminatory actions again LGBTQ+ individuals can include microaggressions against an individual. The goal for mathematics teachers is to make conscious the subconscious heteronormativity, gender normativity, and LGBTQ+-phobias that are hard to avoid in our society. For some newer to these ideas, this means that you must work hard to eliminate your prejudicial thinking or any discriminatory actions you might take because of them. We will look at more specifics for how this plays out in our classrooms and next turn to some of the basic theories on sexuality to inform our thinking more deeply.

Theories and concepts on sexual identity

To start reviewing the deeper ideas at play in theory of sexuality, we review the considerations of famed social theorist Michel Foucault. Foucault's work contributes significantly to the poststructuralist perspective on social life. He examined a variety of topics, from mental health to knowledge to imprisonment, by consistently describing how what is at one historical moment perceived to be normal and "a fact" is actually a constructed reality that represents a range of historical confluences, from politics to economics and other social relations. There exist "discourses" and "regimes of truth," what we might take as given, for granted, always-having-been-true ideas that actually shape our relations to one another and significantly impact one's place within a web or flow of societal power. We can conceive of sexuality, gender, and even mathematics as such regimes of truth. Working among a variety of topics, Foucault describes discursive formations, or the processes that herald such discourses as regimes of truth. His work and that of his peers framed much of social theory taking place in the 1990s and beyond, all of which loosely corresponds to a poststructuralist theory.

One of the topics addressed by Foucault is the notion of sexuality in his book first published in English in 1970: *The history of sexuality: An introduction Volume 1* (1990). The book describes the discursive formation of sexuality at the end of the nineteenth century and continuing to the present day. Foucault discusses the relations between an uprising bourgeoisie and capitalist economics and the beginnings of repressed sexuality. Examples include newfound legal and practical considerations claiming the reproductive role of sexuality as paramount.

These newer circumstances invented new identities fitting outside this definition, such as the homosexual, tainted by claims of perversion. In spite of such negative distinctions, we make progress throughout history as, for example, with movement toward LGBTQ+ rights. However, these triumphs exist within such categories and definitions and sometimes may reinforce the superior/inferior binary at the root of the problem. Foucault writes:

> There is no question that the appearance in nineteenth-century psychiatry, jurisprudence, and literature of a whole series of discourses on the species and subspecies of homosexuality, inversion, pederasty, and 'psychic hermaphrodism' made possible a strong advance of social controls into this area of 'perversity;' but is also made possible the formation of a 'reverse' discourse: homosexuality began to speak in its own behalf, to demand that its legitimacy or 'naturality' be acknowledged, often in the same vocabulary, using the same categories by which it was medically disqualified.
>
> *(p. 101)*

Take, for example, contemporary progress of LGBTQ+ movements that have successfully pushed against the original claims that homosexuality is perverse or disorderly. Some of the fuel for these successes has been to articulate queerness as biologically based. However, Foucault claims, as above, that such success came as the result of claims that further define homosexuality as an "other." Typically, poststructuralism argues that advancement comes with more fluid understandings of binaries rather than strict definitions that relate in some way to a "provable" scientific fact such as genetics. Similarly, Foucault would argue that marriage equality acclimates the "other" into a sanitized, repressive sexuality. This is not to say that these conversations and successes are worthless in fighting the regime of truth that has cast non-heterosexual activity as inferior and disorderly.

In setting up our connections to the mathematics classroom, Christopher Dubbs synthesizes these tensions into the categories of assimilation and liberation movements within LGBTQ+ progress. "The assimilationist groups did (and continue to) fight for social acceptance; often their methods involve minimizing differences and emphasizing sameness through essentialization" (p. 1043), whereas liberationists attempt to identify themselves on their own terms and without seeking to compromise their identities or differences from societal expectations or norms. Dubbs shares this for the relevant approaches we see in public education's efforts to make progress for LGBTQ+ students. On the one hand, there are softer approaches, such as increasing representation of LGBTQ+ people in the curriculum; on the other, we can deepen critique of curriculum and classroom practices that subtly reinforce heteronormative expectations and assert its homophobic standpoint.

Mathematics teachers should also be informed by reading the LGBTQ+ community's history. Stein's (2012) *Rethinking the gay and lesbian movement* provides carefully studied historical development of progress with respect to rights for minoritized sexual identities. The historical context of homophobia and oppression of minoritized sexual identities, along with the steady and hard-fought progress via social movements from the 1950s to today, helps us know the LGBTQ+ community and its numerous struggles and triumphs in more depth. Stein begins the history with the prevalence and reception of same-sex affection and sex among African, indigenous, and European people at the start of the 1500s in North America. He continues through major stages of gay and lesbian history in the United States, including what was known as homophile activism in the early twentieth century, to the 1970s gay and lesbian civil rights era that resulted in a conservative backlash later in the decade, to the activism during the 1980s' AIDs crisis, and finally to the ways that the lesbian and gay activism more broadly focused on LGBTQ+ activism through to today. One of Stein's themes throughout detailing this history is the entanglement of different sexual and gender identities as these movements came about and made progress. In some cases, these would cause tensions and differences of opinion but most generally worked together for progress that removed terrible laws, such as those that criminalized same-sex sex, and generally allowed for individuals and communities to assert their sexual and gender identities, including steps against discrimination based on sexuality and gender expression, still a work in progress.

LGBTQ+ students in public education

Similar to the histories of LGBTQ+ communities, LGBTQ+ youth in classrooms and schools continue to face significant hardships despite recent progress and success in some areas. In this section, we will review some of the current policies affecting LGBTQ+ youth in schools as well as learn from among the numerous articles, books, and research studies about minoritized sexuality and gender identities in schools.

As one way to consider the LGBTQ+ student experience in schools, we can review policies that specifically address their realities. In the United States, the majority of policies enacted to help support LGBTQ+ students are at the state level and with wide variance throughout the country. One state, such as New Jersey for example, may have very supportive policies, whereas its neighbor, Pennsylvania, does not. As LGBTQ+ in schools is a current issue with consistent changes, we as mathematics teachers should be up on the current policies and practices in our area. These are readily available to find on the internet such as with the website "Movement Advancement Project." At the time of this writing, their "Safe Schools Laws" map shows that 23 states of the United States have laws prohibiting bullying based on sexual and gender identity and 2 states have teacher or school codes that prohibits bullying LGBTQ+ students. Sadly, 20 states have no laws or codes preventing LGBTQ+ bullying, and, even worse, 2 states have a law prohibiting school districts from adding policies to include LGBTQ+ youth in their anti-bullying policies (Movement Advancement Project, 2022). The range demonstrates that some states are proactive in addressing LGBTQ+ bullying issues while some are proactive in perpetuating LGBTQ+ bullying.

Bullying of LGBTQ+ youth is perhaps the most studied and clearest problem facing LGBTQ+ students in our schools. One well-known school study comes from Pascoe's (2012) *Dude, you're a fag: Masculinity and sexuality in high school*. Although this research includes broader themes of sexuality and masculinity beyond LGBTQ+ youth, Pascoe's careful work paints an all-too-common picture of the complexities of heternormativity and homophobia in today's schools including the consistent bullying that LGBTQ+ youth face daily. Tragic results of bullying are LGBTQ+ youth suicide and self-harm rates, what education researcher Boni Wozolek (2017) refers to as the "school-to-coffin pipeline." Drawing on previous research and their own, she and two LGBTQ+ teens describe what is termed an "epidemic" by many as follows:

> Thoughts of self-harm and suicide do not develop within an isolated vacuum that begins and ends at home. These thoughts often produce events that permeate the classrooms and the corridors daily in schools. In addition to more frequently contemplating self-harm or suicide, LGBTQ youth often experience anxiety and fear for their physical and emotional safety at school. Unfortunately, encounters with cruelty for queer youth are firmly rooted in sociocultural values that allow homophobia to continue as an acceptable norm of bias and hatred in schools and communities. In light of these and other such factors, the physical cuts and emotional scars that tend to pervade LGBTQ students' ways of being often begin in the spaces and places they should feel most safe at school.
>
> *(Wozolek, Wootton, & Demlow, 2017, p. 392)*

Indicating another measure of relative progress for LGTBQ+ youth in schools, we can consider the number of states that have enacted anti-discrimination legislation. These efforts prohibit LGBTQ+ youth from being denied access to equal opportunities to learn, sports, and extracurricular activities and access to school facilities appropriate to their gender or sexual identity. One hot-button issue you have likely seen in the news are some states that have banned transgender students from using the bathroom that corresponds to their gender identity and/or providing gender neutral facilities. Sadly, three states currently have this ban in place at the time of this writing. Thankfully, 18 states have anti-discrimination policies for LGBTQ+ students in place and a few are in progress but the 24 states that have none yet indicate that we have quite more work to enact better policies for LGBTQ+ inclusion in our public schools. LGBTQ+ students need to have these policies in place to affirm their identities in the schools they attend, minimizing as much as possible the negative impact that their marginalized identities will have in full participation of school life. Mathematics teachers can keep up-to date on these policy efforts with the website on anti-discrimination legislation at Movement Advancement Project (2022).

Greater detail and more thorough explanation of LGBTQ+ student experiences and school policies can be found in Cris Mayo's (2014) *LGBTQ Youth and Education: Policies and practices*. Mayo provides careful review of theory and appropriate language practices regarding LGBTQ+ youth, of the realities of homophobia and transphobia in schools, and of reconsidering the most talked-about LGBTQ+ issues, such as by reframing bullying as a form of sexual harassment. Mayo also draws attention to LGBTQ-headed households and the discrimination faced by them and their children in schools, tensions with our efforts for better LGBTQ+ work and religious perspectives that we might have in our school communities, and several important practices that teachers and school leaders can do as important LGBTQ+ work. These include increasing LGBTQ+ representation, its people, communities, and histories, into the curriculum. Mayo also details the effectiveness and obstacles faced in establishing and sustaining Gay-Straight Alliances and Day of Silence and other programming that increase visibility, support, and awareness of LGBTQ+ identities and issues in school communities. Mayo covers the consistent attempts to block progress for minoritized sexual and gender identities, such as the many variations of policies attempting to exclude LGBTQ+ content from the curriculum. Recently, Florida's state legislation made headlines by passing its Parental Rights in Education Act (2022), also known as the "Don't Say Gay Bill," which prohibits any classroom instruction on sexual orientation or gender identity in the first four years of elementary school.

LGBTQ+ students in mathematics and STEM classrooms

Mathematics education research and its practitioners have only begun to look carefully at the lived experiences of LGBTQ+ students in our classrooms. Although the published research and practice in this area is small, what we will review here provides a strong momentum for addressing these issues. We as a mathematics teaching community can do better by engaging with scholarship and practitioner perspectives on queer mathematics students. Themes from these perspectives discussed in this section include the negative experiences of LGBTQ+ people in math and STEM classrooms, the lack of representation and outright heteronormativity and homophobia of LGBTQ+ in curriculum, and the imperative for what some refer to as a "queer turn in mathematics education."

Kersey and Voigt (2021) specifically researched LGBTQ+ students in mathematics and STEM classrooms at the undergraduate level. Their qualitative studies focused on transgender and gender non-conforming students as well as minoritized sexual identities in STEM classrooms. Some themes in their findings include the lack of acceptance that queer students face by the STEM discipline, be it their professors or peers. As one example, the researchers noted a gender fluid student's "psychology and cognitive stress induced by being in STEM spaces while presenting in gender-nonconforming ways" (p. 747). Ultimately, this student changed their major to an area in humanities that would more readily accept and integrity their gender identity. Related to this student's coming-out process, the STEM LGBTQ+ students in this research noted the need to find peers and teachers who were open and receptive to their minoritized sexual and gender identities. This indicates the prevalence of several faculty and peers who are not understanding of sexual and gender diversity, and while it is the case that teachers in other content areas are similarly intolerant, the research does suggest that STEM teachers are more intolerant, a major concern for our field. Kersey and Voigt's suggest that greater faculty and peer understanding of LGBTQ+ students will reduce discriminatory language and microaggressions that affect LGBTQ+ students and increase implementation of queer-inclusive curriculum.

On the notion of curriculum, we as mathematics teachers can audit our classroom environments and materials for their inclusion of LGBTQ+ communities as well as for their heteronormative and homophobic contents. Two publications provide strong examples as inspiration and some guidance on this important work for mathematics teachers. As a research study, Parise (2021) used critical discourse analysis to unpack the underlying messages of heteronormativity and gender binaries in high school statistics textbooks. There were multiple types of content, for example references to binary-only genders, associations between genders and job types, and heteronormative assumptions about marriage relationships. Through a careful process, Parise determined that one typical gendered theme is that females/women/girls were portrayed as mothers, whereas males/men/boys were portrayed as athletes. This both reinforces gender binaries (excluding transgender and non-binary people and communities) and gender norms. Another theme that Paris found is the consistent assumption that one's gender is the same as one's sex. The textbooks worked with male/female statistics repeatedly throughout and conflated the idea that one's sex assignment at birth was the same thing as their gender identity. Unfortunately, this provides repeated messaging of harm to our students, especially those that are transgender.

A final theme from Parise's work with statistics curriculum is the consistent assumption of heterosexuality in relationships. This included that all stock photos were of heterosexual couples throughout with no same-sex couples and problem scenarios that consistently reference heterosexual attractions and attitudes. In one of these,

> the beginning of the problem states that the statistics student 'wonders if tall women tend to date taller men than do short women.' Each ordered pair represents a woman and her corresponding 'man' who is taller than she (with one exception where the partners are of equal height). Engaging in a problem based on that stereotype communicates its acceptances and reinforces it for students.

(p. 776)

Similarly, in an article written specifically for mathematics teachers, mathematics education researcher Laurie Rubel (2016) thoroughly explains the pitfalls of so-called classic problems used in mathematics classrooms. Rubel details the impacts of story problems with marriage scenarios, including the "stable marriage problem" that is typically instructed via a simulation in the classroom. Students are split into male and female categories, and if in the male group, asked to list those in the female group in order of preference for marriage and vice versa. Mathematics instructors don't realize that there are several other, far less harmful, scenarios that can be utilized for the mathematics of stable pairings (such as college admissions processes). When she experienced instruction of this problem, Rubel writes of the harm this way:

> 'Just focus on the mathematics,' I was told, even though I was being handed a pink card and thereby being placed in a particular location on a gender binary. Not only that, but heteronormativity was being reinforced with the statement that, in this model, all women have to want to marry men… Beyond making mathematics seem incompatible with realities of people's lives, by imposing binary structures on gender and requiring heterosexuality, this activity and others like it demonstrate how mathematics can be used to reinscribe already-oppressive formats.
>
> *(p. 438)*

Thus far the LGBTQ+ mathematics education literature reveals to us that LGBTQ+ students are marginalized in our classrooms because teachers and peers present a level of intolerance and because the curriculum lacks adequate LGBTQ+ representation. The curriculum also displays heteronormativity and confirms gender binaries and norms, thereby reinforcing only dominant gender and sexual identity expression to further marginalize the minoritized gender and sexual identities that are present in our classrooms.

The final theme in the LGBTQ+ mathematics education literature is the urgency for our community to act in addressing these issues, what some are terming as the needed "queer turn" in mathematics education. Dubbs (2016) announced the need for this turn and for our community to center the experience of marginalized queer students. He reviews the important work done in these areas but calls our attention to just how much more work is needed to be done. On curricular improvements, Dubbs points us to the work of Rands (2009) as he describes two approaches, one much more essential to the work ahead. The first and less impactful approach, which Rands calls the "add-queers-and-stir," signifies the attempts to include LGBTQ+ representation into our curriculum but with little major changes to our approaches. As an assimilationist project, this approach would limit discussion of the fluidity of gender and sexual identities; as Dubb writes, they would "generally have a goal of inclusion in homonormative ways" (p. 1045), meaning that these additions reinforce what we expect of LGBTQ+ individuals and communities rather than embrace acceptance of diversity within all minoritized gender and sexual identities. Rands points us to "mathematical inqueery" as stronger examples of disrupting the heteronormativity and homophobia in our mathematics curriculum. The examples include several that provide an intersectional focus, such as this fourth-grade investigation of a local newspaper that relates LGBTQ+ identities to race:

> Students might examine local newspapers and explore the following questions: In the local newspaper with the highest number of readers, what percentage of the articles mention queer people or issues? Of these, what percentage portray queer people in positive ways, negative ways, and mixed positive and negative ways? What percentage of articles mention people of color? Of these, what percentage portray people of color in positive ways, negative ways, and mixed positive and negative ways? What percentage of the articles specifically mention queer people of color? Of these, what percentage portray queer people of color in positive ways, negative ways, and mixed positive and negative ways? [Afterwards,] students could examine the same questions in a local queer newspaper and local newspapers targeting communities of color. What stories do these different percentages tell? Students could write letters to the editors of each of these newspapers sharing these findings and pointing out who is being included and excluded in each newspaper. These letters could serve as counter-narratives challenging both racism and heteronormativity.
>
> *(p. 186)*

In these examples, Rands begins providing answers to Dubbs' call for more intense curricular work in our classrooms but we need to continue this work consistently in our classrooms. For more inspiration, Rands includes several other examples across grade levels as well as integrating the work with other facets of human diversity, like social class.

Moore (2021) provides a literature review of LGBTQ+ mathematics education research with similar urgency to Dubbs, noting the limitations of "papers that adopt the position that diversity can be addressed by incorporating lesbian or gay characters into word problems" because this approach "pathologizes queerness rather than liberating it" (p. 667). Moore writes that the intersection of LGBTQ+ and mathematical identity is a "new borderland to inhabit," one that requires "willingness … to embrace entirely new ways of thinking; the ability to critique one's own preconceived beliefs and understandings of words, notions, and concepts; and the admission of how each of us acts as oppressor" (p. 668). Although he writes specifically for a call to mathematics researchers' capacity for self-examination of how their work either reinforces or interrupts heteronormativity, homophobia, and transphobia, the same is clearly true for all of us as mathematics teachers teaching our students in classrooms.

In reviewing the newly emerging field of queer research in mathematics education, we see several new departure points we must take in our classrooms. This includes increasing representation of the queer community in our curriculum, addressing LGBTQ+ bullying and discrimination, and developing mathematical tasks that foreground gender and sexual identities. As an area of new and active inquiry, we also must stay active in reading new developments about greater equity for LGBTQ+ students in mathematics education.

Activities and prompts for discussion:

1. Locate the current laws in your state or area related to LGBTQ+ students in schools. How proactive are they in addressing school-related problems for your queer students? These include anti-discrimination and anti-bullying laws that hopefully make specific mention of LGBTQ+ students; other proactive policies in some areas include active support for LGBTQ+ youth homelessness, sex education inclusive of LGBTQ+ topics, and the restriction of gay and gender identity conversion therapies. What are some of the policies in your region that you are pleased to see are in effect, if any, and how can they support your work with LGBTQ+ mathematics students? What are some that you'd like to see changed, and what are existing community organizations that you can support with your volunteer hours to make changes in these policies?
2. Discuss with a partner examples of homophobia that you have witnessed in a school setting. What took place? What was your role at the time (teacher/ student), and did you address the homophobia that you witnessed? If you saw a similar action take place again, what would you do this time?

Tips for the classroom:

*   When you meet a new class or group of students, take time for introductions and request that students provide their preferred pronouns. Normalize for your students the practice that we share pronouns when we meet new people.
*   Intervene when homophobic language is used by students, teaching your students the harm that derogatory language causes and that you aim to provide an inclusive environment for all learners.
*   Provide discreet support for students that you suspect are facing challenges related to LGBTQ+ issues, such as by being bullied by peers or facing intolerances at home. This can look like a conversation after class in which you simply say, "I've noticed [this happening] and I want you to know about [this resource]," such as a school counselor that you trust can really help with what the student is going through. Make sure to be careful about sharing

any information that you have about a student with their family as they may not feel safe to discuss this at home.

- Develop a good relationship with the faculty advisor of your school's Gay-Straight Alliance. Talk with them about the idea of adding an LGBTQ+ STEM event to their calendar. If your school doesn't have a GSA or other group with a specific LGBTQ+ focus, think about starting one.
- Queer your curriculum, first start by increasing representation of LGBTQ+ people, communities, and histories into your learning environment and mathematical tasks. Include posters and add biographical and historical information to lesson plans. Deepen this practice by auditing your mathematical tasks in the curriculum. Remove and modify those that reinforce heteronormativity and gender binaries. Finally, take significant time to develop extended tasks that focus on intersectional identities inclusive of LGBTQ+ communities, using the examples from Rands (2009) as inspiration.

TABLE 7.1 Important terms and concepts in this chapter

LGBTQ+	Lesbian/Gay/Bisexual/Transgender/Queer+; One among several options of acronyms that names the community of minoritized sexuality and gender identity; the + indicates the multiple other identities not present in the acronym, such as asexual, intersex, aromantic, and pansexual
Queer	An umbrella term to represent the community of minoritized sexuality and gender identities, similar to LGBTQ+ but with noted differences in use and meaning
Heteronormativity	Society's consistent messaging of the dominant and preferred sexual identity as opposite-gender sexual and romantic attraction; appears in all institutions including media, government, religion, and schools
Homophobia	Intolerance, bias, prejudice, and discrimination of same-sex sexual and romantic attractions
Transphobia	Intolerance, bias, prejudice, and discrimination against transgender individuals
Human sexuality	One's physical, spiritual, and emotional attractions for intimate interpersonal relating; encompasses romantic and sexual attractions including desires and actions
LGBTQ+ inclusive curriculum	Representing LGTBQ+ people, communities, and histories in mathematics and other curricula
Queering mathematics classrooms	Addressing and eliminating homophobic/transphobic and heteronormative interpersonal interactions in classrooms; enacting mathematical tasks with deep engagement of LGBTQ+ representation and issues (beyond "token inclusion" of LGBTQ+ identities in word problems)

Further reading

Dubbs, C. (2016). A queer turn in mathematics education research: Centering the experience of marginalized queer students. In Wood, M.B., Turner, E.E., Civil, M., & Eli, J.A. (Eds.) *Proceedings of the 38th Annual Meeting of the North American Chapter of the International Group for the Psychology of Mathematics Education*. The University of Arizona, pp. 1041–1048.

Kersey, E., & Voigt, M. (2021). Finding community and overcoming barriers: Experiences of queer and transgender postsecondary students in mathematics and other STEM fields. *Mathematics Education Research Journal 33*(4): 733–756.

Mayo, Cris. (2014). *LGBTQ youth and education: Policies and practices*. Routledge.

Moore, A. (2021). Queer identity and theory intersections in mathematics education: A theoretical literature review. *Mathematics Education Research Journal 33*(4): 651–687.

Movement Advancement Project. (2022). Equality maps: Safe school laws. https://www.lgbtmap.org/equality-maps/safe_school_laws

Parise, M. (2021). Gender, sex, and heteronormativity in high school statistics textbooks. *Mathematics Education Research Journal 33*(4): 757–785.

Pascoe, C. (2012). *Dude, you're a fag: Masculinity and sexuality in high school* (2nd edition). University of California.

Rands, K. (2009). Mathematical inqu(ee)ry: Beyond 'Add-queers-and-stir' elementary mathematics education. *Sex Education 9*(2): 181–191.

Rubel, L. (2016). Speaking up and speaking out about gender in mathemathics. *Mathematics Teacher 109*(6): 435–439.

Stein, M. (2012). *Rethinking the gay and lesbian movement*. Routledge.

Wozolek, B., Wootton, L., & Demlow, A. (2017). The school-to-coffin pipeline: Queer youth, suicide, and living the in-between. *Cultural Studies Critical Methodologies 17*(5): 392–398.

8
DISSOLVING ABILITY BINARIES IN MATHEMATICS EDUCATION

From special education law to disability studies

Much attention has been paid to mathematics students with disabilities, and the particular efforts and approaches have changed over time. Since the 1970s, special education laws have prioritized efforts like response to intervention (RTI), a key development that greatly increased opportunities for special education students to learn mathematics. We will also consider new traditions including disability studies and DisCrit, focusing on the ways segregation has limited opportunities given the societal impressions that disabled students are less than or inferior to "abled" students. For example, research demonstrates that disabled students are often not provided opportunities to engage in higher-order thinking in mathematics classrooms. Mathematics teachers who confront such ableism directly in their classroom will think carefully about their efforts to differentiate such opportunities across the ability spectrum, and we will consider practical examples of what this looks like.

Rethinking special education with disability studies and DisCrit

Special education law provides a typical starting point for discussing students with disabilities in public schools. In the United States, the Americans with Disabilities Act (ADA) of 1990 ensured that all educational institutions provide accommodations so that students with disabilities have access to facilities, programs, and activities. More broadly, the ADA prevents discrimination based on disability in jobs, housing, public facilities, and other areas. Also in the same year, the Individuals with Disabilities Education Act (IDEA) replaced previous federal legislation for special education student populations, this time clearly emphasizing that all students with disabilities be provided with a free and appropriate public education. IDEA names several special education practices for those familiar

DOI: 10.4324/9781003322566-8

with US public education including the Individualized Education Program (IEP) and the Least Restrictive Environment.

These educational policies and practices aim to eliminate discrimination by promoting integration in schools for students with disabilities. Historically, students with physical and cognitive disabilities were often segregated or excluded from the educational opportunities provided to other students. Significant gains in opportunity for students with disabilities have been made by these laws and practices; yet there is more work to be done in eliminating discrimination and providing full educational opportunities for students with disabilities. The clearest path for continued progress comes from the field of disability studies, a critical perspective that challenges long-held assumptions about people with disabilities and is now increasingly applied to education. After reviewing some key ideas in disability studies, we will also look to a related field, DisCrit, and how this helps us to think about race and disability in the classroom.

As with other chapters in this book, we begin discussing disability with the most advanced thinking on this facet of human diversity before moving to what is known about disability and public education. There are a variety of introductory texts that nicely introduce the burgeoning field of Disability Studies. One of these, by Berger (2013), presents the relevant concepts of disability studies including its movement beyond what is known as the "medical model" of disability. In this classic model, cognitive and physical disabilities are diagnosed at the individual level through a clinical setting. This approach casts a negative perspective on an individual's capacity or limitations as compared to what other individuals can and cannot do. As Berger writes, the rise of the medical model meant that people with disabilities "were now deemed worthy of medical diagnosis and treatment and viewed more benevolently. But benevolence may breed pity, and the pitied are still stigmatized as less than full human beings" (p. 2).

On the contrary, disability studies refocuses our thinking by suggesting that the nature of disability is not located in an individual. Berger describes that while an impairment "refers to a biological or physiological condition that entails the loss of physical, sensory, or cognitive function" (p. 6), a disability is where one's impairment meets social life and society's expectations for what a person can do.

> For instance, people who use a wheelchair for mobility due to a physical impairment may only be socially disabled if the buildings to which they require access are architecturally inaccessible. Otherwise, there may be nothing about the impairment that would prevent them from participating fully in the education, occupational, and other institutional activities of society… Moreover, people with disabilities often experience prejudice and discrimination comparable to what is experienced by people of color and other minority groups, and they are therefore socially marginalized and disadvantaged in similar ways.
>
> *(p. 7)*

In this way, people with disabilities are viewed as inferior or, as Berger puts it, "not normal." Disability studies reveals society's pathologizing and pitying of people with disabilities and calls us to act toward eliminating the notion that people with disabilities are inferior. An individual's impairment results in a disability only because society has limited how it allows people to access and experience the social world. The ways in which we organize social living and experience are the causes of disability for individuals, not the other way around.

Just like other facets of human diversity, ability and disability are socially constructed concepts determined by the norms and patterns developed in society. Society minoritizes individuals with disabilities and discriminates against them; this is called ableism. Ableism is readily seen everywhere you look once you start to notice it. An easy place to find it is in common language practices. As Berger describes:

> Disability studies asks us to become more aware of the words and phrases we may use, sometimes intentionally and sometimes unintentionally, that demean people with disabilities (such as 'gimp,' 'spastic,' or 'retard'), including metaphors that conflate physical impairment with mental impairment (such as 'turning a blind eye' or 'turning a deaf ear')… In contrast, disability studies often uses 'people first' language, referring to 'people with disabilities to emphasize the person rather than the disability. However it is also common practice … to use the term 'disabled people' to highlight disability as an affirmative identity, not one to be ashamed of, that identifies the common cause of a particular political constituency.
>
> *(p. 5)*

Ableist language practices are as commonplace as or even more prevalent than homophobic language practices. One step that mathematics teachers can take is identifying ableist language in their own speaking and eliminating it, and also identifying it in students' language and other members of their school community as a step toward raising awareness about these discriminatory actions.

Another complementary theory on disability is Dis/ability Critical Race Studies (DisCrit) which relates the disability studies movement with Critical Race Theory. As an intersectional theory, DisCrit engages race and disability together to layer individual social identities that reveal new understandings. Among DisCrit's tenets, it posits the idea that the social constructions of disability and race work historically and currently in tandem to reinforce notions of inferiority for both people of color and people with disabilities:

> Race and ability are socially constructed in tandem, the perception of race 'informing' the potential abilities of a student and the abilities 'informing' the perceived race. Simultaneously, DisCrit rejects what Crenshaw has called the vulgarization of social construction, where critics claim that if race is considered a social construction, then it should be seen as

insignificant and be ignored. In other words, while recognizing the social construction of particular identity markers, such as race and ability, DisCrit acknowledges that categories hold profound significance in people's lives, both in the present and historically. The error, however, made by those who make a false distinction between race as a social construction and dis/ability as a biological fact, distinguishing dis/ability from aspects of identity that are seen as culturally determined 'differences,' continues to justify the segregation and marginalization of students who are considered dis/abled from their 'normal' peers… This phenomenon is particularly true for students of color with dis/ability labels who are more likely to be segregated that their white peers with the same dis/ability label.

(Annamma, Connor, & Ferri, 2013, p. 13)

DisCrit emerges from education scholarship with powerful descriptions of the entanglement of race and ability constructs as manifested in public schools. Although this scholarship exists in education, I placed DisCrit in this first section in this disability chapter as a complement to disability studies because of its powerful theoretical attention to intersectionality for students of color with disabilities. We continue in the next section with additional findings from research on students with disabilities in public education that are informed by both disability studies and DisCrit, before moving on to particular findings for mathematics education.

Disability studies in schools: Universal design in learning and three waves of inclusion

Since the 1990s, thinking critically about disability using ideas from disability studies has become a strong perspective for advancing policy and practices for students with disabilities in public schools. The disability studies approach to education is often referred to as Disability Studies in Education (DSE) and emphasizes the socially constructed nature of disability within the school setting. A clear focus for DSE are the policies and practices related to Special Education services throughout public education, and DSE recognizes the advancements made for students with disabilities since ADA and IDEA as well as the significant work needed to be done to fully address what is still pervasive ableism in schools. In this section, we will look first at some of DSE's foundational questions posed for educators, next at a method called Universal Design for Learning (UDL), and third at what has been called the "three waves of inclusive practice." Each will help us to think about important questions about our practices in mathematics classrooms.

Baglieri, Valle, Connor, and Gallagher (2011) provide a history of the emergence of DSE in the 1990s at a critical time when the special education field was stuck in debates over "incremental" or "reconceptualist" approaches to bettering the education of students with disabilities. By asserting a perspective on the socially constructed nature of disability, DSE helped the field break through these conversations with clarity on major issues in special education like continued segregation and tracking of students with disabilities. DSE presents clear questions about access for students with disabilities, calling all educators to consider first how ability and disability are socially constructed and at the individual level are fluid; in other words, we cannot be held to strict assumptions about what our students in and out of special education can or cannot do. Second, we need to observe and address how our learning environments and student-student and teacher-student interactions are the source of disabilities for students. Finally, we must query "how pull-out, tracking, or containment practices both mark individuals as disabled and/or limit their access to curriculum and learning" (p. 272). DSE also calls into question the prevalent advances for education of students with disabilities, typically the steps toward inclusion like coteaching and modifying and accommodating instruction:

> Many if not most, of us in DSE struggle with the unintended consequences of these approaches even though we share the inclusive ethos that animates them. Our concern centers on the way that the concepts of accommodation and modification contribute to the separation or partitioning of types of students as 'special needs' as opposed to 'typical' or 'general education.' One way that the concepts of accommodation and modification contribute to isolation is that they presume the 'rightness' of a normal curriculum and set of teaching practices. If these are unsuccessful for some students, it is the struggling child who is deemed problematic, rather than our curricular choices and pedagogical practices. The students who do not experience success become *the problem* to be accommodated and the ones for whom modifications are needed, which stigmatizes the individual…Subsequently, we labor away at trying to fix or remediate the students rather than altering the teaching and learning conditions in the classroom.
>
> *(p. 272)*

An approach to curriculum design, lesson planning, and instructional delivery that DSE finds moving us more toward full integration for students with disabilities is Universal Design for Learning (UDL). Authors Hitchcock, Meyer, Rose, and Jackson (2002) on UDL reject the notion of a homogenous classroom supposedly in which the majority of students have similar needs; this mindset causes us to accommodate or modify our teaching for students with disabilities when in fact every student in a group has key differences that matter for their learning. The starting place is designing instruction that includes a range of options for

> accessing, using, and engaging with learning materials – recognizing that no single option will work for all students, UDL shifts the burden for

reducing obstacles in the curriculum away from special educators and the students themselves and leads to the development of a flexible curriculum that can support all learners more effectively.

(p. 9)

UDL provides recommended approaches for four aspects of the lesson design process: goals, materials, methods, and assessment (pp. 11–14). For all, flexibility and multiplicity are the overarching suggestions. For goals, UDL reminds educators to state these learner goals broadly and with multiple pathways for success. This stands in contrast to lesson plan literature, typically steering us to learning objectives that then need to be modified for students with disabilities. Additionally, goals should reflect curricular standards. UDL is not an effort to "water-down" the curriculum; it is the process of opening it up to a greater number of learners. Materials can include multiple types of media from text to voice to images and addresses the range of students we have in classrooms. Having multiple materials allows students to choose the content that matches their needs best while attaining the learning goals. In a mathematics lesson on right triangles, Hitchcock et al. (2002) suggest that multiple pedagogic methods can be

> multiple examples of right triangles in different orientations and sizes; an animation of the right triangle morphing into a [non-right] triangle, with voice and on-screen text to highlight the difference; links to reviews on the characteristics of triangles and right triangles; links to pages that students can go to on their own for review or enrichment on the subject.
>
> *(p. 13)*

Finally, UDL reminds us that barriers to access exist commonly in our assessment practices. We must remain steadfast in focusing on the ways we elicit our students' newly acquired knowledges and skills, making sure we are attending to how our assessment practices allow for open access by our students with disabilities.

Another helpful consideration emerging from DSE scholars are what some (e.g., Shogren & Wehmeyer, 2014) term the "three generations" or "waves of inclusive practices." Inclusion is the idea to provide students with disabilities access to the same curriculum as other students and Shogren and Wehmeyer (2014) noted that this began first with *where* students with disabilities learn. The first policies and practices to occur closed separate school facilities for students with disabilities and integrated them into the neighborhood public schools. The second wave of inclusion continues a focus on *where* but adds *how* to teach students with disabilities with key developments in strategies like coteaching for students with disabilities in mainstream classes. Coteaching is a standard practice in many public schools today in which a subject-specific teacher (or general teacher at the elementary level) is partnered with a special educator who has specific training in pedagogies for students with disabilities and additional knowledge of the students with disabilities in the given class. Coteaching is successful in moving toward inclusion, although DSE research reveals how typically cotaught classrooms continue to signify an "inferior" classroom of student learners because many other classrooms, those without students with disabilities, are taught by only one teacher.

The final wave of inclusion includes the *where* and *how* and also adds the *what* of the curriculum. Driven especially by DSE research as well as the popularity of the new UDL, inclusion now includes thinking about what kind of knowledges and skills are expected for students with disabilities to learn. Inclusion that helps to integrate students into the same space is one good thing that removes social stigma but these stigmas can only perpetuate if students with disabilities are held to routinely inferior learning outcomes. While it can be true that a student's particular impairment may result in a different expected learning outcome, the consistent pattern of lowering expectations for students with disabilities is too commonly seen and renders our schools and classrooms as sites that reinforce, even teach, the ableism existing throughout society. As we will see in the next section, some of the mathematics education research drawn from DSE squarely takes on the issue of *what* math curriculum is made accessible to students with disabilities.

Mathematics classrooms for students with disabilities

Disability studies in education (DSE) has recently been applied directly to mathematics education policy and practice. DSE has revealed that our work has integrated new developments aligning with special education law but we are far from providing the full opportunities for students with disabilities, thus continuing to perpetuate ableism in our society. At present, mathematics education has advanced significantly in "accommodating" students with learning disabilities, and we will start by reviewing typical approaches, like the currently in-favor response to intervention model. A teacher guide that uses this model was put out by the US Institute of Education Sciences: *Assisting students struggling with mathematics: Response to intervention (RtI) for elementary and middle school students* (2009). It provides how-to assistance in the early detection of and accommodation for learning disabilities as they relate to mathematics instruction. Situated within the efforts of the 2004 reauthorization of the Individuals with Disabilities Act, the guide's goal is to

> provide suggestions for assessing students' mathematical abilities and implementing mathematics interventions within an RtI framework, in a way that reflects the best evidence on effective practices in mathematics interventions. RtI begins with high-quality instruction and universal screening for all students. Whereas high-quality instruction seeks to

prevent mathematics difficulties, screening allows for early detection of difficulties if they emerge. Intensive interventions are then provided to support students in need of assistance with mathematics learning. Student responses to intervention are measured to determine whether they have made adequate progress and (1) no longer need intervention, (2) continue to need some intervention, or (3) need more intensive intervention.

(p. 4)

An example of the recommendations contained in this report is the emphasis on visual representations to augment understanding of mathematical ideas.

Although the intended efforts of RTI support students with disabilities, unintended consequences can result. Ferri (2010) critiques RTI as a tactic that reinforces exclusionary practices in school systems. She claims that RTI is the most recent iteration of special education reforms that objectify students with special needs:

Once deemed eligible for special education, students are assumed to be 'fundamentally different' from their non-disabled peers. Disability labels, therefore, function as a discursively produced system of social othering that creates divisions between students who are considered normal and regular and those who are seen as deficient and disordered.

(p. 1)

Unfortunately, the RTI mathematics guide from the Institute of Education Sciences commits these faults and we need to look instead to efforts within mathematics education that are informed more fully by DSE.

More generally, mathematics education research focusing on students with disabilities typically continues these mediocre efforts that do not fully embrace the DSE perspective. Tan, Lambert, Padilla, and Wieman (2019) reviewed this research to find that most research perpetuated deficit-based perspectives on students with intellectual disabilities in the mathematics classroom. They continue to locate a student's disability at the individual level and needing to be accommodated, fixed, or otherwise handled. The authors of this review note a growing number of researchers that mention the social nature of disability but these maintain an individualized approach characteristic of most research efforts. Of the 50 researcher articles reviewed by Tan et al., only 4 utilized a framing consistent with DSE's distinction that disability is a social construct and that mathematics classroom environments and teaching strategies, materials, etc., are the problem, not the student with the disability. Their review of this research proves very important for the directions we need to take as mathematics teachers committed to the most advanced thinking and practice for our students with disabilities.

Tan et al. write about the important differences in this type of research that fully utilizes the perspectives of DSE:

In contrast to the dominant forms of research in mathematics education involving students with intellectual disabilities that focuses on direct forms of instruction and basic mathematics skills development, undermining conceptual construction and understanding of mathematics, the studies in this category presume, to a greater extent, that students with an intellectual disability are mathematics thinkers and doers, capable of a range of mathematics engagement.

(p. 8)

The studies point out that mathematics teachers and researchers typically interpret a student's impairments as the need to accommodate or modify learning goals in ways that reduce the teaching and learning to only procedural skills and without inclusion of mathematical reasoning, application, or other aspects of mathematics. The studies in mathematics education that are grounded by DSE reject that students with disabilities should be limited in such ways.

Yeh, Ellis, and Mahmood (2020) propose that mathematics teachers of students with disabilities prioritize teaching practices that actively disrupt an inferior label or track for students with disabilities and are culturally relevant for their students. They also remind us of the mind-body connection and that physical manifestations are an important aspect of mathematical behavior. In their research, Yeh et al. describe a mathematics teacher of elementary students with disabilities. At this stage in their learning, the students are continuing to work on one-to-one correspondence (ability to answer "how many?" when given a set) yet the teacher assembles the group of students and works on division concepts with them. To a traditional mathematics teacher, the students are "not ready" for this work because they have not mastered prior concepts and also, as the research above shows, typically, the students would not be afforded opportunities to explore division but instead just be shown how to do it:

They would not have been thought capable of engaging with 'higher-level' mathematics concepts like partitive division. Ms. Huerta intentionally does not reduce the curriculum's level of complexity. For her, mathematics learning is not linear and does not follow hierarchically organized progressions. Instead, mathematics is experienced as a social and relational activity involving her students. The division concept is embodied first in the multiple hands that lay out the pieces into rows of fives to make tens, then in the distribution into individual cups, next in the voices that called forth, questioned, and revised the count, and finally in the formal abstraction as Ms. Huerta wrote up on the whiteboard the two ways to represent the situation: the total number of students, 4, indicates that the total number of M&Ms, 24, is 6 times as many as the number of M&Ms each student will receive. Mathematical meaning and matter are bound and woven together through complex embodiment of the students with each other, the material resources, and the verbal count leading to exploration of the ways in which one makes meaning of division. This also [developed] ... other mathematical ideas: meaning of numeral, ordinal counting, mental manipulation, and spatial reasoning.

(p. 7)

Put simply, the students were given rich tasks with multiple points of entry for differing learners to access and work together on socially. The division task was meaningful because they were asked to share candy equally among the group. Ultimately, students engaged the concept and demonstrated understanding of division through body movements and verbal and non-verbal communications. They were not instructed on procedural steps for division; they were given the opportunity to construct a mathematical idea together through a meaningful activity. This topic was on grade level and typically would not have been taught to the group because they did not have "mastery" of prior grade-level material. Mathematics teachers can similarly reject our stubborn commitments to prior knowledge that limit student opportunities for learning.

As a final support for our discussion on students with disabilities, Lambert (2020) provides clear guidance on designing UDL mathematics lessons for practicing mathematics teachers. She describes the typical barriers faced by students with disabilities and key elements of UDL for mathematics; together these serve as a nice summary of many of the points made throughout this chapter. For barriers, Lambert includes the "limited avenues for learning mathematics" with a reference to the standard use only of lectures and textbooks for materials in our instruction; our insistent "focus on speed and memorization" that especially confronts students with particular cognitive impairments; the pattern that we limit "connections to concepts" for students with disabilities, reserving problem-solving and connections driven instruction only for other students; and the sad fact that "mathematics remains the only subject with a related anxiety – math anxiety – suggesting that emotions raised by traditional instruction are a powerful barrier to many students in mathematics" (p. 7).

Moving to the promises of UDL for mathematics instruction, Lambert suggests we "create safe classrooms" with multiple opportunities to access material and try free from judgment; to "offer relevance and choice" that engage our students with disabilities' cultures and provide options for social and individual learning; to "focus on core ideas" rather than assigning massive sets of practice problems; use "multiple representations" to provide access for all learners, moving far beyond typical visual and text-based representations to also include sound and touch; and to focus on "developing strategic, expert mathematicians" who can reason, problem-solve, and apply mathematics to real-world situations through mathematical modeling. Each of these guidelines includes references to research in mathematics classrooms illustrating what much better instruction for students with disabilities can look like.

Although DSE frameworks for mathematics classrooms is relatively new, our community will continue to see more and more engagement in this area with greater dissemination of practices. As you consult research and other publications on mathematical instruction of students with disabilities, make sure to engage the author's perspective on disability. If they suggest a medical model, the impairment as a problem caused by the student and needing correction, look elsewhere for more informed research. The suggested practices contained in such conversations will continue to perpetuate ableism in your classroom and throughout our society. On the contrary, work with UDL and DSE aligned publications as you think carefully about what flexible instructional practices you will provide all learners in your classroom.

Activities and prompts for discussion

1. Together with a partner, reflect on ableist language practices that you encounter on a regular basis. You can perform an internet search on these practices that will reveal a rather long list of words, some of which you may not have realized are ableist. Commit to eliminating these words from your own speaking and to teaching about ableist words to someone else. Keep a journal on these efforts and check in with your partner two weeks later on your progress.

2. A mathematics teacher might assume that geometry concepts and ideas are not possible to teach to learners with visual impairments; however, acting on this thought is a direct action of ableism because it renders curricular material inaccessible to some students. Select a geometry standard in a grade level that you teach. Discussion question: How would you approach developing meaning for the topic's concepts and skills for learners who have visual impairment? Mathematics teachers need to approach all topics with a UDL approach in practice and mindset.

Classroom tips:

- Address ableist language in your classroom community by eliminating it in your speaking and educating your students about ableism and how it comes through in the words we use.
- Use UDL frameworks for lesson design by collecting a range of materials, using a variety of methods, and using a variety of assessment strategies. Consult guidelines on UDL specific to mathematics teaching and learning, such as Lambert (2020).
- Differentiate the learning experience for students flexibly, for example by changing student groupings in your classroom often. Do not set rigid student groupings that label students with disabilities or suggest that they are completing inferior levels of work.

- Avoid teaching students with disabilities only procedural mathematics. Make sure that reasoning, problem-solving, mathematical modeling, and thinking and doing mathematics in other ways are made highly accessible for your students with disabilities.

TABLE 8.1 Important terms and concepts in this chapter

Impairment	A biological or physical condition that results in loss of physical, sensory, or cognitive function
Disability	Socially constructed barriers or limitations to an individual that are the result of one's impairment
Disability studies	A movement in research rejecting deficit-based medical models of disability; prioritizes disability as an affirming identity and that people with disabilities are not inferior; applied to education as Disability Studies in Education (DSE)
DisCrit	A theoretical and intersectional concept emphasizing the interplay between one's racial/ethnic identity and ability identity; demonstrates confluence of ableism and racism operating simultaneously for students at the individual level and systemic levels
Universal Design for Learning (UDL)	A lesson planning approach for full access to curriculum for all learners, especially those with disabilities; emphasizes flexibility and multiplicity in all aspects of instruction including goals, materials, methods, and assessments
Response to Intervention (RTI)	A typical framing for special education mathematics learners; although has provided some progress, critiqued by DSE for its capacity to reinforce labels and segregation of students with disabilities
Inclusion	Historically, students with disabilities have been separated from general education students; inclusion addresses this segregation by reintroducing students with disabilities into general education schools and now providing greater access to the curriculum

Further reading

Annamma, S., Connor, D., & Ferri, B. (2013). Dis/ability critical race studies (DisCrit): Theorizing at the intersections of race and dis/ability. *Race Ethnicity and Education* 16(1): 1–31.

Baglieri, S., Valle, J., Connor, D., & Gallagher, D. (2011). Disability studies in education: The need for a plurality of perspectives on disability. *Remedial and Special Education 32* (4): 267–278.

Berger, R.J. (2013). *Introducing disability studies.* Lynne Rienner Publishers.

Ferri, B. (2010). Undermining inclusion? A critical reading of response to intervention. *International Journal of Inclusive Education 16* (8): 863–880.

Gersten, R., Beckmann, S., Clarke, B., Foegen, A., Marsh, L., Star, J.R., & Witzel, B. (2009). *Assisting students struggling with mathematics: Response to Intervention (RtI) for elementary and middle schools* (NCEE 2009-4060). Washington, DC: National Center for Education Evaluation and Regional Assistance, Institute of Education Sciences, U.S. Department of Education. Available at http://ies.ed.gov/ncee/wwc/publications/practiceguides/.

Hitchcock, C., Meyer, A., Rose, D., & Jackson, R. (2002). Providing new access to the general curriculum: Universal design for learning. *Teaching Exceptional Children 35*(2): 8–17.

Lambert, R. (2020). *Increasing access to universally designed mathematics classrooms.* Policy Analysis for California Education. Available at https://files.eric.ed.gov/fulltext/ED605096.pdf

Shogren, K., & Wehmeyer, M. (2014). Using the core concepts framework to understand policy, practice and research related to the three generations of inclusive practices. *Inclusion 2*: 237–247.

Tan, P., Lambert, R., Padilla, A., & Wieman, R. (2019). A disability studies in mathematics education review of intellectual disabilities: Directions for future inquiry and practice. *Journal of Mathematical Behavior 54*. doi:10.1016/j.jmathb.2018.09.001

Yeh, C., Ellis, M., & Mahmood, D. (2020). From the margin to the center: A framework for rehumanizing mathematics education for students with dis/abilities. *Journal of Mathematical Behavior 58*. doi:10.1016/j.jmathb.2020.100758

9

LANGUAGE DIVERSITY AS AN ASSET

Emergent bilinguals in the mathematics classroom

Dominant discourses and practices for linguistic diversity in education and mathematics classrooms label some students as "English as a Second Language," what many consider to be a label fraught with a deficit mindset. In this chapter, we will reframe our thinking about this student group by thinking not that those in it *don't have* English but instead see that they *do have* two language practices, even when their English is early in development. Many working in this area thus refer to this student group as "emergent bilinguals" because it reminds us that another language proficiency beyond English is a good thing, an asset to our classrooms and to our society. Whereas you will see most literature refer to the group as ESL or English Learners, throughout this chapter I will refer to our students representing linguistic diversity as emergent bilinguals.

We begin the chapter by looking at the landscape of language diversity in the United States and a discussion of policies and conversations indicating why emergent bilinguals are a minoritized group of students in our schools. We'll next review advanced thinking on linguistic diversity including concepts like translanguaging and raciolinguistic ideologies which are beginning to be applied to mathematics classrooms. We'll conclude with key considerations and practices for working with emergent bilinguals in our mathematics classrooms. This will include how to locate resources to provide language-specific supports for students and how to engage English learners in collaborative work with peers that fosters growth in learning mathematics.

Diversities of emergent bilinguals

Mathematics teachers can embrace language diversity more readily in their classrooms. Often, many of us tend to think that mathematics is a "universal language"

DOI: 10.4324/9781003322566-9

that transcends spoken languages. What difference could it make when our students come to us speaking different languages? Language diversity matters a great deal to our classroom spaces just as the other facets of human diversity do. For one, there are the labeled identities by schools indicating a language minoritized student, labels like "ESL student" or "English learner"; unfortunately, these labels often signify inferiority or a deficit to peers and teachers. On the flip side, a student's language practices beyond English are of great importance to their identity. A student's language practices are deeply personal; within these practices are the many important, emotional connections with loved ones and their community. Language and culture are tightly interwoven as well. Moreover, a student's language practices are assets to the mathematics classroom for their capacities to communicate mathematical knowledge and practice between peers and teacher to student. Whether a student's language practices include English, another language, or both, mathematics teachers must engage a student's entire linguistic repertoire for the best learning to occur. As we will see in the research later in the chapter, this develops students mathematically as well as with their English practices and practices in other languages.

Emergent bilinguals in our classrooms are not all the same. In the United States, there are several languages representing this group of students. The National Center on Education Statistics (2022) demonstrates this diversity with their official numbers from 2019: 5.1 million emergent bilingual students in US public schools, an increase of over 1% of the whole student population in the ten years prior. Home language varies among the group, with Spanish as about 76% of all emergent bilinguals, Arabic as about 3%, English and Chinese each at 2%, and other home languages including Vietnamese, Portuguese, Russian, Haitian and Haitian Creole, Hmong, and Korean. English as a home language for emergent bilinguals represents some students who have multilingual, English dominant home experiences as well as children adopted by English-speaking parents who have language experiences outside of English.

Other ways that emergent bilinguals are different are their race/ethnicities, socioeconomic backgrounds, and origins. Many assume that all emergent bilinguals are born outside of the United States; however, many emergent bilinguals are born in the United States as well. Both emergent bilinguals that are born in the United States and those that are immigrants to it have connections to other countries. Emergent bilinguals have connections to or origins in several places, including Asia, Latin America, Europe, Africa, and Pacific Islands as well as Native American and other indigenous populations. For each of these groups, it is important to notice the diversities existing within. We should avoid generalized thinking about groups of students, such as "all my Spanish speakers are Mexican American." Emergent bilinguals with Spanish as their home language represent a variety of cultures and countries including Mexico, El Salvador, Guatemala, Dominican Republic, Cuba, and Columbia and also Puerto Rico, a US territory. Each country has unique cultural features and linguistic practices that differ from each other so these differences in our students are important to recognize.

Emergent bilingual students in the United States face significant challenges that can be related to their language backgrounds. They are more likely to be economically disadvantaged compared to other US students. Among all emergent bilinguals, Asian and white English learners are more likely to have more favorable economic circumstances compared to BIPOC English learners (National Academy of Sciences, 2017, p. 80). Many emergent bilinguals are undocumented immigrants or have a parent or family member who is undocumented. This further exacerbates hardships that our students and their families will face. "Recent estimates suggest that more than half of all English learners have an undocumented immigrant parent" (p. 88), and this causes stress due to the threat of deportation, potentially fracturing a family group. Additionally, our students who are undocumented immigrants will face difficulties as they progress through the education system and apply for colleges and financial aid.

These hardships cause significant barriers for opportunity in school and in our mathematics classrooms. They vary for each individual student based on their complementing identity factors beyond language including race, dis/ability, and social class. However, one barrier is present for most all emergent bilinguals: the expectation of assimilation to the United States through demonstrated proficiency in English. In many cases, this results in segregation into English learner education programs until a level of proficiency is achieved; in others, it means reduced expectations by teachers with a poor education as the result. Several negative approaches to emergent bilinguals, even at the policy level, have been harmful. Assimilationist policies, such as removal of bilingual education programs, have been demonstrated as "subtractive schooling" (Valenzuela, 1999), meaning that with such policies school adds nothing to emergent bilinguals and only subtracts their home language and culture. Repeated studies have shown that assimilationist practices, English-only programs, and the like do not effectively teach emergent bilinguals today. The promising research for emergent bilinguals comes from those educational spaces that develop these students as they emerge into their full language practice using two or more languages. We next take a look at the core concept, translanguaging, that grounds these spaces.

Essential concepts for emergent bilingual education: translanguaging and raciolinguistic ideologies

In response to the overarching and failing approach of teaching English only to emergent bilinguals, educational linguists like Ofelia García promote teaching students in ways that fully engage and develop their linguistic repertoire. Emergent bilinguals come to classrooms wanting to communicate and engage with others in ways that they can and often will move fluidly between distinct languages without consciously thinking or realizing that they are doing so. She and others in the field describe these language practices as translanguaging:

> Translanguaging is the act performed by bilinguals of accessing different linguistic features or various modes of what are described as autonomous languages, in order to maximize communicative potential. It is an approach to bilingualism that is centered, not on languages as has often been the case, but on the practices of bilinguals that are readily observable in order to make sense of their multilingual worlds.
>
> *(García, 2009, p. 140)*

Such a view of emergent bilinguals or language minority students in our classrooms positions them as eager communicators drawing on multiple strands, rather than students who *need to have* English skills.

In applying translanguaging to teaching and learning, García describes the pedagogical implications of translanguaging as having two principles: social justice and social practice:

> The social justice principle values the strength of bilingual students and communities, and builds on their language practices. It enables the creation of learning contexts that are not threatening to the students' identities, but that builds multiplicities of language uses and linguistic identities, while maintaining academic rigor and upholding high expectations. Another important element of this principle has to do with advocating for the linguistic human rights of students and for assessment that includes the languaging of bilingual students. The social practice principle places learning as a result of collaborative social practices in which students try out ideas and actions, and thus socially construct their learning. Learning is seen as occurring through doing. Translanguaging among students, especially in linguistically heterogeneous collaborative groups, becomes the way in which students try out their ideas and actions and thus, learn and develop literacy practices.
>
> *(p. 153)*

These principles apply to mathematics classrooms and are consistent with reform mathematics' emphasis on collaborative learning that fosters students to negotiate meaning. With these in mind, mathematics teachers shift their thinking from language minoritized students as having deficits toward students with advantage.

As we will see when we look to practical applications of translanguaging in the mathematics classroom, the approach is similar to opening access to curriculum like we discussed for students with disabilities. Increased use of home language materials and encouragement of students to communicate with each other and with you using whatever language comes to mind is paramount for a translanguaging space in mathematics instruction. It is important to note that you as the mathematics teacher need not be proficient in the home languages of your students but at the same time must be open to learning about them and seeking out helpful resources. Many textbooks now include Spanish language supports but these will not help your speakers of other languages. As well, these supports can be limited. As you identify the languages spoken in your classroom, seek out resources in your community including proficient bilinguals who might be able to support you. For example, as you search for YouTube materials in other languages, an adult who is bilingual can provide you with a rough idea of the video's merits so you know what you are providing for your students. There are more practical suggestions especially when you are limited in your experience with a student's home language, and we will look at these in the final section of this chapter.

Having an open mind toward your students' linguistic practices and encouraging them to communicate with all in their linguistic repertoire is a first step to mitigating the harmful status quo of public education for emergent bilinguals. Mathematics teachers also need to be mindful of the ways in which emergent bilinguals are further minoritized and excluded for reasons that entangle with other identities that layer on to individual students in our classrooms. For example, mathematics teachers can remain free of judgment about a student or family member's immigration status and instead understand their student's transnational communities from a place of empathy and support. One area of increased understanding about emergent bilinguals are the ways in which language minoritized students are limited in opportunities to learn because their language practice and racial identities intersect as a double negative within the school system. This is a concept called raciolinguistic ideologies and is pervasive in public education throughout the United States.

Flores and Rosa (2015) suggest with their concept *raciolinguistic ideologies* that certain emergent bilingual students in public schools, particularly BIPOC students, are not seen fully for their English language practices when in public schools. With a strong emphasis on standard American English as a goal, even in many bilingual programs, BIPOC emergent bilingual students are often seen as not proficient in English even when this is not the case. This can explain, for example, why many students can remain in a school system's "ESL program" for seven years or longer. This occurs when individual students are competent in speaking in an English dialect other than standard American English. Teacher recommendations and standardized placements continue to label a student as not speaking English *appropriately*; however, they are speaking with fluency within their certain community of BIPOC English speakers. The concept also explains why heritage language speakers (US-born emergent bilinguals) can also be labeled as lacking English proficiency. In simple terms, raciolinguistic ideologies enacts inferior stigmatizing and segregation when the school system perceives a student's race and ethnicity primarily via their linguistic practices. Mathematics teachers can recognize these actions in their own practices and within their school systems by asking themselves: "Am I (or are we) making judgments about this student because we perceive a need to support their English proficiency or because we are trying to erase their racial identity?" In this way, raciolinguistic ideologies is an intersectional focus that highlights how society perceives an individual's linguistic practice as embedded within their racial identity.

With their concept of raciolinguistic ideologies, Flores and Rosa bring in other areas of linguistic diversity that are equally present in US public schools, such as Ebonics/African American English (AAE). Black and African American students who speak AAE are not identified as English learners in the school systems yet experience a similar hardship with respect to language practices in the school environment. They are routinely dismissed for speaking AAE in schools and given correction to speak standard American English at every turn. In this way, Black and African American students are viewed by schools as being deficient English and language users when in reality they are competent in AAE. Similar to a translanguaging and bilingual education approach, Lisa Delpit has advanced teaching Black and African American students to communicate with AAE in the classroom and also to be taught standard American English as the language of power. Students are taught to recognize the times when to "codeswitch" between the two ways of speaking. In this way, students' home and community language practices are affirmed before and alongside the learning of standard American English.

Emergent bilinguals in the mathematics classroom

As with other chapters, as you engage in new research on emergent bilinguals in the mathematics classroom, look for advanced theoretical understanding including using forward-thinking language and concepts like translanguaging or raciolinguistic ideologies. The review of research and discussion I include here are those that use an explicit or implied translanguaging perspective that views students' home languages as assets rather than deficits. Unfortunately, a lot of research on mathematics education and emergent bilinguals contains deficit framing and should be considered with caution; most of the research published here is published in venues focusing on emergent bilingual pedagogies rather than mathematics education. The research and publications below include full integration of translanguaging into mathematics spaces, to demonstrate what is possible, as well as beginner strategies and "how-to's" for mathematics teachers who are getting started.

Garza (2017) provides a case study of a middle school mathematics teacher who fully embraces translanguaging in his teaching. The school's bilingual program technically required students be taught mathematics in Spanish, but the teacher engaged both languages fully as a means to develop their mathematical literacy more deeply. The students in the classroom were at this point proficient bilinguals, with speaking and reading abilities in both English and Spanish. Although mathematics could likely have been taught successfully in either language, the teacher knew that utilizing the full linguistic repertoire for his students would increase mathematical understanding even more. As an ideal situation, the teacher was also bilingual in English and Spanish but even more so was his understanding of his students' two languages as assets for him to draw out in the learning process. Garza writes:

> Mr. León did not show a preference of one language over the other. Instead, he made use of his linguistic abilities as one linguistic system... He promoted an inclusive and flexible mathematics space that celebrates bilingualism—a space where he and his students were able to draw features from their formal and functional linguistic resources. As proficient bilinguals, Mr. León's students used this space to creatively use their entire linguistic repertoire...While using translanguaging discourses, Mr. León promoted mutual engagement when performing mathematical tasks. In doing so, he and his students developed a joint enterprise in which active mathematical negotiations took place.
>
> *(Garza, 2017, pp. 150–151)*

The examples from Garza's observations indicate how reform mathematics teaching, emphasizing meaning making and reasoning, was enhanced because the teacher was using the full linguistic repertoire of his students. This stands in stark contrast to mathematics teachers who teach only in English, viewing their students as having language deficits to be corrected. Such a perspective extremely limits the potential for student learning and likely will only allow students to learn low-level mathematical procedures, if anything at all.

Garza's case study is complemented in the research by extensive studies like Moschkovich (2015). This research on emergent bilinguals in mathematics classrooms aligns strongly with reform mathematics teaching because it emphasizes the need for higher-level thinking in mathematics instruction for emergent bilinguals. Moschkovich thus pushes back against the overwhelming focus on vocabulary instruction for emergent bilinguals. Typically, "academic language" is separated from developing a student's mathematical proficiency and

> such a separation can make [English learners] seem more deficient than they might actually be, since they may not be able to express their mathematical ideas through language, but may still be engaged in correct mathematical thinking and participate in mathematical practices that are less language intensive, for example using objects or drawings to show a result, finding regularity in data, or using gestures to illustrate a mathematical concept.
>
> *(p. 44)*

Moschkovich suggests three parts to developing an emergent bilingual's full mathematical literacy: mathematical proficiency, mathematical practice, and mathematical discourse. The first includes fluency in mathematical procedures and also ability to reason mathematically and have conceptual understanding; the second, focus on practice, is for Moschkovich a social experience of peer-to-peer and student-teacher mathematical activity; and the third is the ability to communicate mathematical ideas through a variety of means. Mathematics teachers should emphasize and develop all three for their emergent bilingual students. Doing so, especially for the third focus on discourse, means encouraging students to communicate using their full linguistic repertoire of home language and language of instruction. And, as the first focus on proficiency suggests, increasing access for emergent bilinguals means engaging the full range of mathematical practice well beyond low-level mathematical procedures. All too often, emergent bilinguals are offered only a procedural mathematics because this seems to be the simplest to convey across language barriers.

Moving to research in curricular materials, de Araujo and Smith (2022) provide a review of mathematics education supplements that aim to support emergent bilinguals. Their research clearly asserts that mathematics teachers must view their students' language backgrounds as assets for developing mathematical knowledge. Unfortunately, their research findings indicate that the language supports in typical textbooks fall far short in what is needed. First, these supports tended to portray emergent bilinguals as needing the same type of support, mostly reinforcing academic language skills and with little connection to a student's full linguistic repertoire. The supports typically provide additional remediation in skills practice, confirming the idea that emergent bilinguals are thought to only be able to do tasks with low cognitive demand. The research suggests to mathematics teachers that we should be cautious in utilizing textbooks' efforts to support our emergent bilinguals. These will tend to reinforce deficit framings of our students and we will need to be inventive in identifying methods to engage our student's full linguistic repertoire as an asset to learning mathematics.

At this point, some of the strongest practitioner guides and examples of translanguaging for mathematics classrooms comes from collaborative projects focused on emergent bilingual work across all content areas rather than ones that specifically focus on mathematics instruction. One of these is Hesson, Seltzer, and Woodley (2014), a translanguaging guide for educators written for the City University of New York and New York State Initiative on Emergent Bilinguals. This center was developed by Ofelia García and others who are key figures in promoting translanguaging for emergent bilinguals in today's classrooms. This helpful guide provides clear descriptions of the motivations and goals of translanguaging with an opening set of frequently asked questions. For example, they respond to the frequent issue brought up about English-only teaching for emergent bilinguals as follows:

> If the classroom teacher speaks only English, children must be given the opportunity to make sense of the new language through their own home language practices. This means that teachers must set up collaborative

structures in classrooms where children from the same language group may be able to discuss the tasks at hand in their home languages, even if they are being asked to produce the text in English. Even though the teacher does not speak the languages of the children in the classroom, she must create a translanguaging space so that all children make meaning of the lesson being taught and cognitively engage with the material at hand.

(p. 10)

Further, general tips are provided in their opening discussions, yet what is most specific are the example unit plans across content areas that provide "Translanguaging How-to's" throughout.

One of the unit plan examples in their guide is a middle school mathematics unit on proportion, scale drawing, and other aspects of geometry. The unit plan teaches the content by focusing on monuments:

Students are architects and activists in mathematically designing a monument of their choice. Students will begin by learning about real monuments and their dimensions, then individually or in groups, create 2-D scale drawings to represent real-world structures. With multilingual readings, students mathematically explore diverse monuments including the Statue of Liberty, a bilingual monument for Veterans in San Diego, the African Burial Ground National Monument, and the multilingual Wall for Peace in Paris. With these, and with other examples they have found themselves, students will create and solve real-world word problems using mathematical operations of rations, proportions, multiplication, area, surface area, and volume. After learning about existing monuments, their purposes and construction, students will geometrically plan their own new monument. It is important for students to understand the purpose of a monument and put heart and thought into creating something that is meaningful to each individual student in response to a person, place, event, or cause.

(p. 77)

Selecting this theme for the unit allows for rich opportunities in translanguaging mathematics classroom. Students are able to reason mathematically and apply new understandings to the real world, thereby doing mathematics at a higher level than typical for emergent bilinguals. The cultural themes contained in the unit's exploration of monuments allows for rich exploration in a multilingual setting. For example, the teacher can locate materials in the home language about monuments as well as point students in directions to find more of these themselves.

Throughout the unit, there are several various types of how-tos for support-ing the translanguaging space. These can serve as inspiration for use in the differ-ent content areas of mathematics curriculum. Many of the examples encourage students to write in their home language for writing prompts throughout a unit. Another consistent practice is grouping students by home language so they can discuss their progress and questions together using their full linguistic repertoire. One tendency for some teachers is thinking to pair a strong English speaker with an emergent bilingual in the hopes that the English speaker will pull them along. However, this can be a negative approach when the English speaker does not have facility in the other student's home language; often, they resort to doing all the work for both students, and the emergent bilingual's progress suffers as a result. Grouping students in shared language proficiencies is typically a better approach. Flexible grouping and the encouragement of students to utilize their full suite of language practices will lead to the greatest amount of learning in mathematics.

Another consistent practice stressed by the examples in the translanguaging mathematics unit is utilizing a multiplicity of materials. Home language materials are often available for our textbooks, and several additional resources are available if you make the commitment to look for them. Internet video hosting websites like YouTube allow you to search for videos in different languages. These can be tremendous supports for your emergent bilinguals; you can choose to have them watch it before you teach a lesson or during, depending on how you think it might help your students to connect with the material best. Additional resources include reaching out to your community to find guest speakers who can come to the classroom and speak in your students' home languages on topics related to the unit.

In reviewing what we know about translanguaging mathematics classrooms, we aim to provide an environment for our students to use their full linguistic repertoire. We must commit to learning and seeking new information about our content in their home language as much as possible and allow our students to work through material by talking to each other, and us, using the languages that are available to them. As a primary interruption in our practice, we must stop reducing emergent bilinguals' mathematical work to procedural mathematics and vocabulary development. Just like all our learners, our emergent bilinguals will flourish when we provide opportunities for mathematical reasoning, mathemati-cal modeling, and critically thinking with mathematics.

Prompts and activities for discussion:

1. Learn about the linguistic diversity in your local community and region. One place to start is with identifying the languages present in your state or region first. The Migration Policy Institute publishes information for many states

in the United States with specific demographics of English learner populations: https://www.migrationpolicy.org/sites/default/files/publications/EL-factsheet2018-California_Final.pdf. Take a look at the languages in your region to help you locate some of the local communities nearer to you and possibly some community and cultural sites that could serve as linguistic resources for you as you work with emergent bilinguals in your classroom. Also talk with your school communities about the linguistic resources in the community that you have found or that others know about to build greater understanding across your school.

2. Select a mathematics unit in your curriculum and locate the learning supports made available for your emergent bilinguals. Evaluate their usefulness; likely they will only support vocabulary development and basic skills remediation. Using "translanguaging how-to's" as your guide, develop three specific supports for your emergent bilinguals that stress developing your student's higher cognitive work in mathematics, such as by emphasizing reasoning or application.

Classroom tips:

- Group emergent bilinguals who speak the same home languages together and allow them to speak often as they work together, clarifying meaning for the content and advancing understanding together. It is okay if you do not understand their home language much or at all but you should engage with gesturing and other non-verbal cues, as well as reinforce the content by speaking English to your students.

- Emphasize emergent bilingual's capacity for higher-level mathematical thinking and doing by designing supports that help students to engage with conceptual understanding of the material and mathematical applications and modeling.

- Continuously locate resources in students' home languages, from videos to print materials to guest speakers. Integrate these into your classroom to foster its development as a translanguaging space.

- Learn your students' home languages as you have time, especially by listening to movies or shows to get the language into your ear and help with your pronunciation. You will be looking up translations for the key words and concepts for each mathematical unit you teach; it is better when you can pronounce these correctly.

- Use technological to your students' advantage, including live translation software for subtitles on the board when you are delivering new content in English as well as translation apps on smartphones or tablets. Recognize your students' proficiencies in reading and speaking in their home language as you use these. For example, live translation into written text may be helpful for some but not all students.

TABLE 9.1 Important terms and concepts in this chapter

Emergent bilinguals	Students who have home languages other than English or who are otherwise multilingual; typically schools determine these students to be "limited" in English proficiency and labeled "English as Second Language," which does not suggest emergent bilinguals' linguistic assets
Translanguaging	Emergent bilinguals communicate with their full linguistic repertoire at all times rather than deliberately switching from one language to the other; mathematics teachers can create translanguaging spaces by providing students with materials in home languages and encouraging students to write and speak in home languages as they learn content
Subtractive schooling	In English-only and other limiting programs, emergent bilinguals are ineffectively taught English and lose knowledge of their home language and culture, thereby a subtractive educational experience
Raciolinguistic ideologies	By emphasizing academic language and standard American English, schools and teachers mischaracterize students' racial or ethnic identities as English language deficiencies; causes issues like extended remediation programs in English for BIPOC students
Linguistic repertoire	Emergent bilinguals and all individuals have multiple ways of communicating through various languages and means; teachers can effectively communicate when a student's entire linguistic repertoire is accessible and engaged

Further reading

de Araujo, Z., & Smith, E. (2022). Examining English language learners' learning needs through the lens of algebra curriculum materials. *Educational Studies in Mathematics* *109*(1): 65–87.

Flores, N., & Rosa, J. (2015). Undoing appropriateness: Raciolinguistic ideologies and language diversity in education. *Harvard Educational Review 85* (2): 149–171.

García, O. (2009). Education, multilingualism, and translanguaging in the 21st century. In Mohanty, Ajit, Panda, Minati, Phillipson, Robert, & Skutnabb-Kangas, Tove (Eds.), *Multilingual education for social justice: Globalising the local*. Orient Blackswan, pp. 128–145.

Garza, A. (2017). A translanguaging mathematical space: Latino/a teenagers using their linguistic repertoire. In Ramirez, P., Faltis, D., & DeJong, E. (Eds.), *Learning from emergent bilingual Latinx learners in K-12: Critical teacher education*. pp. 139–156. Routledge.

Hesson, S., Seltzer, K., & Woodley, H.H. (2014). *Translanguaging in curriculum and instruction: A CUNY-NYSIEB guide for educators*. CUNY-NYSIEB. Available at www.cuny-nysieb.org.

Moschkovich, J. (2015). Academic literacy in mathematics for English learners. *Journal of Mathematical Behavior 40*: 43–62.

National Center for Education Statistics. (2022). English learners in public schools. U.S. Department of Education, Institute of Education Sciences. Available at https://nces.ed.gov/programs/coe/indicator/cgf/english-learners.

Valenzuela, A. (1999). *Subtractive schooling: U.S. Mexican youth and the politics of caring.* SUNY Press.

10
PUTTING IT ALL TOGETHER

Intersectionality revisited, current mathematics education policy, and further avenues for exploration

Each of the preceding six chapters has taken on a facet of human diversity by thoroughly discussing the relevant literature both within and outside of research on mathematics education. This aims to promote a critical practice in mathematics education that engenders inclusion both in content and in its access to it. Here, and by way of conclusion, we will put these isolated discussions together and in further action in three ways. First, we look to the common themes and patterns that emerge as we now have examined multiple facets of human diversity in the mathematics classroom. Second, it is important to situate our work of critically teaching mathematics within the current context. To do so, I provide a brief review of the history and politics of mathematics education, a somewhat-hard-to-swallow tale for critical mathematics teaching. Following this, the third section responds with a review of essential readings for critical mathematics educators. Put together, the three sections of this chapter aim to properly situate the preceding chapters in a more comprehensive and total framework. Although at times these topics will complicate and make our work more difficult, they are necessary discussions that make our objectives more realistic, practical, and steadfast.

Common themes for diversity in mathematics education: Emphasizing intersectionality, multiplicity, language practices, and high-level content

Now that we have worked through several facets of human diversity, you may have noticed several common themes across the recommendations for teaching made in the literature. In this way, we can gather a set of overarching principles for teaching mathematics for all learners and especially those who are typically

DOI: 10.4324/9781003322566-10

marginalized in our spaces. In this section, we gather the themes of intersectionality, multiplicity, language practices, and high-level content as main takeaways from the discussions contained in this book.

Before moving through each of the facets of human diversity in Chapters 4 through 9, we engaged with the concept of intersectionality to make sure that even at the start we knew that it does not help to think of any identity as an isolated aspect on its own. When learning about the ways to better teach mathematics for students across different identities, did you notice how often the literature we engaged spelled out the ways that multiple student identities converge to create layered experiences in our classrooms? In the chapter on social class identity, research was highlighted suggesting that girls from lower-class backgrounds in particular found obstacles for full engagement in the kinds of pedagogies their teachers were using; here, gender and social class were layered to create a unique situation for student learners. As another example, in the chapters on linguistic diversity and ability we learned that students' perceived dis/abilities and language proficiencies can be misconstrued due to their race or ethnic identity.

This theme of intersectionality runs throughout the literature we have engaged in previous chapters. I included discussion of intersectionality early in Chapter 3 to avoid the misconception that each facet of human diversity works independently even though the organization of this book makes it appear that way. Sociologists and theorists of education tend to separate out facets so that we can understand patterns, but they always remain attentive to the overlaying identities and within-group differences that appear. Similarly, in our classrooms we cannot treat any one student identity in isolation but as a collection of layered and multiple identities working together to form our whole classroom of diverse learners with beautiful assets and needed actions we can take as teachers. Reviewing the research on each facet of human diversity as it relates to mathematics teaching will help us to see these assets and know particular actions to take. As you may have noticed, there do appear consistent approaches we can take to help address all facets of human diversity, albeit with careful attention to the specifics needed for each learner in our class.

The first of these consistent approaches to highlight is the need for multiplicity and flexibility in the classroom. We saw this clearly mentioned in the chapter on students with disabilities and again in the chapter on emergent bilinguals. The general approach for mathematics teachers is to provide multiplicity and flexibility with materials, methods, and assessments to increase access to the curriculum for all learners. You can expand on the notion of Universal Design in Learning (UDL) to include all facets of human diversity as you attempt to provide access for all learners in your classroom. In our chapter on race and ethnicity, we considered how our materials and curricula need to be culturally relevant and sustaining. Offering multiple materials will serve the multiple cultures present in your classroom. Also related to this multiplicity and flexibility is keeping your learning social and with flexibility among student groups.

Another theme that came up consistently concerns our classroom environ-ments that we establish through our language practices. LGBTQ+ and students with disabilities face significant hardship in schools due to homophobia and able-ism. Although explicitly racist and sexist language is nowadays taboo, racial and gendered microaggressions are pervasive in public schools as well. Our language practices also include the way we identify students and whether these labels are demeaning to students (in the case of "ESL student") or something making students feel rather uncomfortable (in the case of saying to a class "Ladies and Gentleman"). Mathematics teachers need to be committed to staying up-to-date on the languages pertaining to diversity; they are ever changing, and make every attempt to eliminate bias from communication whenever possible in the classroom.

A final common point to stress from among these particular recommenda-tions with respect to facets of human diversity is the matter of high-level math-ematical content. Research has shown that time and again the marginalized in our classrooms are not provided access to the mathematics that requires them to "think" and "do" at a higher level, as reform mathematics teaching would like us to stress. Instead, labels of inferiority like "ESL" and "Special Education" and underlying prejudices and assumptions of inferiority in mathematical ability (as in the case of girls and women, BIPOC, and LGBTQ+ students) cause so many of our students to miss opportunities for mathematical tasks of higher-cognitive demand. When working with marginalized students, resist any pressures from your teaching peers or your curriculum to "just teach them the skills."

History and politics of mathematics education

Our next discussion that motivates teaching mathematics with a critical orienta-tion comes with a review of the history and politics of mathematics education. The contents of this section piece together from a variety of sources, including mainstream histories of mathematics education within the research community as well as critical perspectives on mathematics education, including my own work in Wolfmeyer (2014), the work of one conservative mathematics educa-tor, Klein (2003), as well as discussions of history and politics of education in general, such as Joel Spring (2000). Simply put, mathematics education is not exactly a friendly space to the critical perspectives discussed throughout this book. Reviewing the history reveals that mathematics education policy directly conflicts with critical goals; as we shall see, military and business interests have been the primary motivators for the development of mathematics education. This history and politics will be focused on the United States; however, similar trends exist across the globe, and we will look at how these have spread through the increasing use of standardized tests.

This history and politics review begins with an assertion of a traditional math-ematics education during World War II, followed by the new math of the Cold

War, yet another traditional, back-to-basics movement in the 1970s, a new-new math movement of the 1990s, the insertion of the standardized testing industry under No Child Left Behind (NCLB), and finally the development and implementation of a national mathematics curriculum called the Common Core. As well, we look at the influence of two international mathematics tests, the Trends in International Math and Science Study (TIMSS) and the Programme for International Student Assessment (PISA). In tracing the histories through a focus on politics and policy, we will make clear a battle over traditional versus reform pedagogy and the primary business and military interests at play in mathematics education.

To begin, first consider the educational stage set by reformers of the 1920s and 1930s. Commonly referred to as the progressive era, several aspects were at play. On the one hand, the progressives aimed to make an efficient, orderly education that streamlined the process and placed people according to the roles society had deemed for them. On the other, educational experiments built off the philosophies of John Dewey, who asserted that education be rooted in experience. In this framing, mathematics was taught through experiences and out of necessity. For example, students might work on a carpentry project and thus be required to learn fractions.

With this progressive era as the backdrop, military leaders during World War II claimed that their soldiers had poor training in basic mathematical skills. The most famous of these, Admiral Nimitz, declared that more specific and direct instruction in both basic and advanced mathematics was needed for naval officers. Such a motivation for a strong US military was well received by the populace given the fight in a so-called good war to defeat fascism in Europe. Nevertheless, this first national conversation clearly situates a national mathematics education within the needs of the US military.

Continuing these military motives, the 1950s Cold War era pushed national mathematics away from traditional, basic skills emphasis and toward what we now call reform mathematics teaching. Although this may seem a surprising about face, the concerns about Sputnik and the Soviet Union's outcompeting in the arms race motivated the United States to develop a curriculum that groomed the nation's "best and the brightest." This translated to mathematics curriculum that taught students early on to think abstractly like mathematicians do, to set them on a path toward advanced mathematics, engineering, and physics. A team of mathematicians, headed by Ed Begle, developed a "new math" curriculum for use in schools. National curriculum is not possible in the United States, due to the separation of powers between federal and state and local governments. However, the federal government was able to have de facto control over the curriculum by providing funds to states and local education agencies which promised to use this new curriculum developed for the military interest (Spring, 2000). Thus, new math found a home in public schools throughout the country.

The 1970s saw a backlash movement against the new math. This was comprised of teachers and parents but spearheaded by mathematicians and professors who were not pleased with the preparation of students arriving in their introductory college classes.

> In 1962, a letter entitled *On The Mathematics Curriculum Of The High School*, signed by 64 prominent mathematicians, was published in the *American Mathematical Monthly* and *The Mathematics Teacher*. The letter criticized New Math and offered some general guidelines and principles for future curricula.
>
> *(Klein, 2003, p. 185)*

Thus, the 1960s new math and 1970s back to basics movement launched an ongoing debate among mathematicians and applied mathematicians commonly known as the math wars. On the one hand, you have the traditionalists who emphasize fact mastery and skills with the logic that students can learn these first and then understand the concepts and theory of mathematics later. On the other hand, the reformists emphasize that mathematics education should mimic the work of mathematicians with a focus on the process of mathematical thinking. At the policy level, thus far the debate over national mathematics curriculum was motivated by military defense and security.

Moving to the next decade, the 1980s Reagan era shifted this focus. With the publication of the policy brief *A nation at risk*, the goals of mathematics education moved toward economic security and dominance. This document detailed the threat against the US economy by the economies of West Germany and Japan. With clear commitments to mathematics and science, these efforts called for national curriculum that would better prepare students for a workforce to compete in the global economy. Continuing in the trajectory set up by new math, the reformists, mostly research mathematics educators, answered this call for a national mathematics curriculum in 1989 with the National Council of Teachers of Mathematics (NCTM) *Standards*. These standards are framed by economic concerns, and more closely orient toward a reform mathematics program emphasizing process over skills and fact mastery. Their updated standards in 2001 somewhat compromise this stance but still remain entrenched in what has been termed a "new-new math." Yet again, these standards were not officially a national curriculum. However, the National Science Foundation supported research and development of standards-aligned curriculum that ultimately found its way into a majority of textbooks used in public classrooms.

Also in 2001, US Congress passed the reauthorization of the Elementary and Secondary Education Act (ESEA) with the title No Child Left Behind. As with each reiteration of ESEA since its origins in the 1960s, NCLB prioritized a fair and just education for all students regardless of gender, socioeconomic status, or race/ethnicity. NCLB's take on this was the requirement that all states have standards and standardized assessments in place. Again, the federal government does not have control over education, but has implemented such national education policy by tying requirements to a state's receipt of federal funding. NCLB clearly situates within concerns over preparing a strong workforce; a lesser-known fact of NCLB is that local districts are required to provide the names and contact information of students to military recruitment officials. Therefore, NCLB continues the theme of commitments to the US economy and military.

Perhaps more influential on mathematics education, NCLB's policy requirement of standardized testing introduced a host of educational businesses (non-profit and for-profit) that would provide standardized testing services. These assessment companies were interested in a unified national curriculum for the sake of efficiency; delivering products to 50 different states was more costly than delivery of one product. In addition, the math wars continued to battle over the new-new math's over-emphasis on process and looked to balance the curriculum with a traditional focus.

Along the lines of a global economic competition, there has existed a global competition among the world's mathematics students. Nations compete against each other for the top math scores among its general student population. Primarily, two tests are viewed this way: the Trends in International Mathematics and Science Study (TIMSS) and the Programme for International Student Assessment (PISA). Both have strong commitments to educating for the economy; TIMSS was at one point sponsored by the World Bank and PISA is sponsored by the Organization for Economic Cooperation and Development (OECD). Putting these points together, performance in mathematics education is seen primarily as an indicator of economic possibility. Ever more and more countries are added to the list of participation in the TIMSS study that parallels the spread of capitalism throughout the globe.

Back to the United States, mathematics educators and mathematicians looked to other countries that outperformed the United States on these tests, as it turns out to justify a more traditional curriculum. Using phrases "internationally benchmarked" and "coherent curriculum," US researchers looked specifically to the national curriculum programs of authoritarian capitalist countries like Singapore and South Korea. These curricula emphasize a highly sequential structure to learning mathematics that the NCTM new-new math standards lacked. Ultimately, these research efforts led to the development of the Common Core State Standards for Mathematics. Also involved were the standardized testing companies so eager in promoting a national curriculum for their profit. And, as usual, these standards were made a de facto national curriculum through the Race to the Top program of 2011, in which states competed for federal funds and were required to adopt the Common Core State Standards for Mathematics (CCSSM) in order to do so.

My own study (Wolfmeyer, 2014) on how CCSSM came to be highlights in great detail the extent to which the economic priorities are in place, as well as the testing and other industries' influence on mathematics education. The majority of individuals and organizations involved in writing national mathematics education policy hold firmly stated commitments to a public education that increases corporate profit. As Joel Spring writes, there is an inherent tension between a public education for the people and a public education for the development of corporate profit. For example, the 1980s fostered a greater relationship between schools and corporate profit, a time when we also witnessed a greater stratification of wealth.

Other industries at play in mathematics education include the information and communications industry, mostly because standardized tests can now be delivered in digital format. The influences on mathematics education are now interested not only in developing these assessments in digital format, but in using digital instruction as well. Major players stand to gain great profits through the activity of public mathematics education, including companies like Microsoft, Apple, Pearson, and ETS. Finally, my analysis of mathematics education policy reveals a continued interest in national mathematics education for the US military.

In the Trump and post-Trump era of US public education, we have also seen increased polarization regarding topics in mathematics education and curriculum. With increased partisanship and corporate media alignments, many confusing messages appear that fuel community responses to equity work in mathematics classrooms. A conservative right wing often characterizes the Common Core as an aggressive attack on states' rights in public education and mischaracterizes the standards as overly progressive and liberal, when in reality they were a compromise with traditional perspectives. Progressive efforts in addressing equity in mathematics education, such as Seattle, Washington's new mathematics education that integrates ethnic studies have been major gains in our policy arena but have faced intense scrutiny by right-wing media outlets. More broadly, these backlashes to equity efforts in public education are omnipresent, such as the significant concerns that the right wing has about applying Critical Race Theory to public education.

Such a history of mathematics education policy in the United States, and its spread across the globe, indicates strong commitments to educating for corporate profit and for strengthening the military. We also see significant opposition to our efforts toward equity in mathematics classrooms. These run counter to education with a critical perspective that would seek to promote democratic goals and personal fulfillment. Certainly, such a context makes our work in teaching mathematics critically more difficult, but it is highly important to recognize these challenges and work toward change. In this vein, the next section's selection of critical points of interest from among mathematics education should provide hope and a sense of solidarity in your work.

Searching for allies: Further avenues for exploring critical work in mathematics education

The majority of resources drawn on within this book, especially in Chapters 4–9, come from the more critical strands of mathematics education. In my efforts to familiarize you with the landscape of critical perspectives on mathematics teaching, I conclude this book with introductions to some further reading and areas for exploration along these lines. As it turns out, many mathematics educators identify with the label Critical Mathematics Education (CME), and some do this kind of work but might not accept the label. My goal here is to sketch a landscape of this work as a response to the current context described in the previous section. I encourage you to seek out these spaces for further exploration and thinking about mathematics teaching and learning.

One recent edited collection on critical work is *Opening the cage: Critique and politics of mathematics education* (2012) by Ole Skovsmose and Brian Greer. It contains chapter essays written by some of the most prominent mathematics educators arguing from a critical perspective. Its introduction, written by Skovsmose and Greer, traces the origins of what they term a "critical mathematics education" and reference several key scholars and concepts. One of these contributions is Skovsmose's work himself: an almost 40-year-long quest to theorize a critical mathematics education which he began in the 1980s and continues today. Another topic Greer and Skovsmose point to is the destabilization of mathematics as exclusively Western, particularly in the work of D'Ambrosio's launch of the field of ethnomathematics. A third strand of critical mathematics education pointed to in the introduction is the specific application of Paulo Freire's praxis to mathematics education, initiated by Marilyn Frankenstein and further developed by Eric Gutstein.

Speaking more broadly about the more critical work in mathematics education, Skovsmose and Greer write:

> The field of mathematics education, in general, has considerably matured, as reflected in the diversification of influential disciplines and related methodologies – broadly speaking, the balancing of technical disciplines by human disciplines such as sociology, sociolinguistics, anthropology, psychoanalysis, and of formal statistical methods by interpretative methods of research and analysis. Within the field, there is heightened cultural and historical awareness, both within and beyond academic mathematics, and an increased acknowledgment of the ubiquity and importance of 'mathematics in action' and the implications for mathematics education, including more curricular prominence for probability, data handling, modeling, and applications. In relation to the political nature of the enterprise, there is greater attention to the relationships between knowledge, education, and power.

(pp. 3–4)

The introductions to thinking about mathematics education critically that have occurred throughout this book are the results of this turn in mathematics education research.

Similar to this chapter's preceding section, Greer and Skovsmose paint a sad picture of present-day mathematics education:

> A great deal of the world's intellectual talent in mathematics (and science) is used in the creation of better ways of killing, subjugating, or surveilling and controlling people, of which current deployment of flightless aircraft, 'drones,' provides a chilling example.
>
> *(p. 5)*

They describe mathematics as cast in "an illusion of certainty" by which

> people abdicate the responsibility of making judgements in complex social situations. [As well,] people and institutions within mainstream mathematics education too often collude with the political establishment by willfully remaining oblivious of the social and political contexts outside their self-constructed cage.
>
> *(p. 5)*

The edited anthology contains several and varied responses to such a context of mathematics, mathematics education and the world. One of these is by Eric Gutstein, in which he announces that "mathematics is a weapon in the struggle." By this, he suggests that mathematics should not be discounted for its relationship to power and destruction, as above, and instead harnessed for its potential to read and write the world. In the chapter, Gutstein documents his research efforts in teaching an urban mathematics classroom where he taught discrete dynamical systems through a unit on HIV/AIDS in the students' community. The students modeled the epidemic and through this negotiated several difficult mathematical concepts and skills as well as challenging social, economic, and political conversations. Moving far beyond the mathematics curriculum that urban students typically learn, the unit also engaged with social theory concepts, such as the intersectionality of social identities (race, class, and gender) that many do not learn until graduate school.

Other essays in the anthology include Alexandre Pais' critique and deep theoretical exploration of the strand in mathematics education research that focuses on equity, arguing that such research reinforces exclusionary practices rather than achieve the goals its rhetoric implies. Marta Civil's work on a critical mathematical education for immigrant students is also included in the anthology, as well as Sikunder Ali Baber's discussion of mathematical education in Pakistan and other international perspectives. In total, the chapters in *Opening the cage* move between critique and possibility, providing a sure understanding of the context in which critical teachers of mathematics find themselves but also the potentials and inspirations toward which we can strive.

As the authors point out, the world of mathematics education research is large and full of variations. Many individuals and organizations commit to an objective of mathematics education aligned to corporate profit and war, at the very least by not taking direct stances against this. Others take steps toward "equity" but do not embrace fully the social theory explained throughout this book, from understanding white supremacy to critiques of capitalist logics. In other words, the field of mathematics education can be daunting for a novice critical teacher of mathematics, and it is important that allies identify the spaces in which critical work is published and communities where the work is promoted. A forthcoming volume, edited by Brian Greer, Ole Skovsmose, and David Kollosche (in production), will no doubt continue these conversations with updated conversations for the present day. These final, subsequent paragraphs point you to the academic journals and communities in which you will find critical orientations to mathematics.

Two international research mathematics education journals do not have explicit orientations to critical work in mathematics but have fostered a space for this work within their objectives. *For the Learning of Mathematics* and *Educational Studies in Mathematics* contain a variety of topics relevant to research in mathematics education, but authors such as those throughout this book regularly publish in these journals. Although browsing these journals will demonstrate that some research fails to acknowledge the social and political dimensions to mathematics education, at least a few articles in most issues will do this. Two journals in the United States also have critical orientations and reputations. These are *The Mathematical Enthusiast* (formerly *The Montana Mathematical Enthusiast*) and *Journal of Urban Mathematics Education*. This is not to suggest that other mainstream journals like the *Journal for Research in Mathematics Education* do not contain critical work. However, as an example, this journal prioritizes articles that contain empirical findings and is less likely to publish articles that push the theoretical boundaries of critical mathematics. Increasingly, you will see that this publication and the others put out by the National Council of Teachers of Mathematics are attending to equity issues directly. Attending professional conferences put out by NCTM will be similarly hit and miss, with several presenters not necessarily using the most advanced thinking on human diversity as applied to the mathematics classroom. The mathematics education conference with the most dedication to our efforts in the sociopolitical turn in mathematics education is international research conference Mathematics Education and Society. Additionally, very important spaces in the United States for your professional development will include the annual conference put on by the Algebra Project. The 2022 conference was titled "We the People – Math Literacy for All."

Finally, with some digging and searching, you will find that work in critical mathematics finds a home outside of spaces dedicated to mathematics education. In the articles from the critical perspective journals like *Educational Studies* and *Critical Education*, you will find several articles (including some of my own) that address equity issues in mathematics education with more critical perspectives. All of this is to suggest that whether you are a reader thirsty for more critical mathematics work or an emerging writer looking for publication venues, you should consider the framing of journals carefully and search in both mathematics education journals as well as critical education journals.

In this concluding chapter, I have attempted to bring together the previous, isolated discussions on mathematics, teaching mathematics, race, class, and gender together by looking at intersectionality, the history and politics of mathematics education, and further avenues for exploring critically teaching mathematics. Intersectionality serves to make our work more complex with a matrix approach to understanding individuals, communities, institutions, and power. In teaching mathematics critically, such an intersectional view privileges the pervasive logics of exploitation and domination that occur throughout our patriarchal, Eurocentric, heteronormative, and ableist culture. A critical look at the history and politics of mathematics education reveals just how much we are up against, yet also is essential knowledge for us to determine where spaces exist to insert our critical orientations. Finally, I have pointed you to some spaces that embrace critical orientations to and critiques of mathematics education, including academic journals that I encourage you to browse and select articles to read.

Thus, my final contribution, and indeed the spirit throughout this book, has been calling you to action. You may find yourself sympathetic to certain arguments contained in the book, and others less so. My goal has been to expose you to the more critical perspectives on mathematics education and resonate with some of them deeply in the hope that you will transfer your empathy for all your learners across the facets of human diversity. I also doubt after our discussions that you can honestly make the claim that mathematics education is objective and agreed upon by all. Mainstream mathematics education might not state its assumptions of corporate power and war, or its orientation as a "white institutional space," or its masculine obsession with rationalism, but these political, economic, and thereby oppressive components exist in it. Critical mathematics educators recognize this and critique it as well as propose and practice alternatives.

And, for those of you that embrace most or many of the critical orientations included throughout, I encourage you to take up this world of teaching mathematics critically in both practice and theory. We need active minds working in communities who educate learners in mathematics. We all can be literate in the debates and controversies on mathematics education, rather than complicit with what is handed to us. As well, we need more minds contributing to the dialogue. Critical orientations to mathematics education have been worked on for at least 40 years, and as such this is a young discourse. The essential spirit of this community is to challenge, to dig deeper, to question. As Skovsmose and Greer remind us, such critique in this community has at once a notion of challenge *as well as* today's crises at heart. Please challenge mathematics education for the sake of justice, for peace, for all living things having more control over decisions that impact them, and for an inhabitable planet with all life at its center, rather than humans. I ask you to, in this spirit, challenge the work in mainstream mathematics education *and also* to challenge the words I have written in this book and the many who engage in teaching mathematics from a critical perspective.

Activities and prompts for your consideration:

1. Discuss with a partner the common themes across Chapters 4–9. What are the examples of the common themes on intersectionality, language practices, higher-cognitive demand, and multiplicity/flexibility? Are there other themes that you noticed across the varying chapters?
2. After reading the section on history and politics of mathematics education, what do you think about a national curriculum for mathematics? Should one exist and, if so, what should its goals and orientations be?
3. Browse articles in the journals mentioned in the third section. Select an article by an author that is new to you and share with your peers.
4. Where do you stand in terms of critically teaching mathematics? Do you agree with some, but not all, of the critiques contained throughout this book? How do you think differently about mathematics education now that you have read this book?

Further reading

Greer, B., Skovsmose, O., & Kollosche, D. (in preparation) *Breaking images: Iconoclastic analyses of mathematics and its education*. A volume in the series "Studies on Mathematics Education and Society" Open Book Publishers.

Klein, D. (2003). A brief history of American K-12 mathematics education in the 20th century. In Royer, James (Ed.), *Mathematical cognition*. Information Age Publishing, pp. 175–259.

Skovsmose, O., & Greer, B. (eds.) (2012). *Opening the cage: Critique and politics of mathematics education*. Sense.

Spring, J. (2000). *The American school: 1642-2000* (5th edition). McGraw-Hill.

Wolfmeyer, M. (2014). *Mathematics education for America? Big business, policy networks, and pedagogy wars*. Routledge.

INDEX